Gender, Health and Information Technology in Context

Health, Technology and Society

Series Editors: **Andrew Webster**, University of York, UK and **Sally Wyatt**, Royal Netherlands Academy of Arts and Sciences, The Netherlands

Titles include:

Ellen Balka, Eileen Green and Flis Henwood (*editors*)
GENDER, HEALTH AND INFORMATION TECHNOLOGY IN CONTEXT

Gerard de Vries and Klasien Horstman (*editors*)
GENETICS FROM LABORATORY TO SOCIETY
Societal Learning as an Alternative to Regulation

Alex Faulkner
MEDICAL TECHNOLOGY INTO HEALTHCARE AND SOCIETY
A Sociology of Devices, Innovation and Governance

Herbert Gottweis, Brian Salter and Catherine Waldby
THE GLOBAL POLITICS OF HUMAN EMBRYONIC STEM CELL SCIENCE
Regenerative Medicine in Transition

Jessica Mesman
MEDICAL INNOVATION AND UNCERTAINTY IN NEONATOLOGY

Nadine Wathen, Sally Wyatt and Roma Harris (*editors*)
MEDIATING HEALTH INFORMATION
The Go-Betweens in a Changing Socio-Technical Landscape

Andrew Webster (*editor*)
NEW TECHNOLOGIES IN HEALTH CARE
Challenge, Change and Innovation

Forthcoming titles include:

John Abraham and Courtney Davis
CHALLENGING PHARMACEUTICAL REGULATION
Innovation and Public Health in Europe and the United States

Health, Technology and Society
Series Standing Order ISBN 978–1–4039–9131–7 hardback
(*outside North America only*)

You can receive future titles in this series as they are published by placing a standing order. Please contact your bookseller or, in case of difficulty, write to us at the address below with your name and address, the title of the series and the ISBN quoted above.

Customer Services Department, Macmillan Distribution Ltd, Houndmills, Basingstoke, Hampshire RG21 6XS, England

Gender, Health and Information Technology in Context

Edited by

Ellen Balka
Simon Fraser University, Canada

Eileen Green
University of Teesside, UK

Flis Henwood
University of Brighton, UK

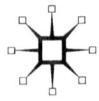

First published 2009 by
PALGRAVE MACMILLAN

Palgrave Macmillan in the UK is an imprint of Macmillan Publishers Limited,
registered in England, company number 785998, of Houndmills, Basingstoke,
Hampshire RG21 6XS.

Palgrave Macmillan in the US is a division of St Martin's Press LLC,
175 Fifth Avenue, New York, NY 10010.

Palgrave Macmillan is the global academic imprint of the above companies
and has companies and representatives throughout the world.

Palgrave® and Macmillan® are registered trademarks in the United States,
the United Kingdom, Europe and other countries.

ISBN-13: 978–0–230–21634–1 hardback

This book is printed on paper suitable for recycling and made from fully
managed and sustained forest sources. Logging, pulping and manufacturing
processes are expected to conform to the environmental regulations of the
country of origin.

A catalogue record for this book is available from the British Library.

A catalog record for this book is available from the Library of Congress.

10   9   8   7   6   5   4   3   2   1
18   17   16   15   14   13   12   11   10   09

Printed and bound in Great Britain by
CPI Antony Rowe, Chippenham and Eastbourne

Contents

List of Tables

List of Figures

Series Editors' Preface

Medicine, health care, and the wider social meaning and management of health are undergoing major changes. In part this reflects developments in science and technology, which enable new forms of diagnosis, treatment, and the delivery of health care. It also reflects changes in the locus of care and burden of responsibility for health. Today, genetics, informatics, imaging and integrative technologies, such as nanotechnology, are redefining our understanding of the body, health, and disease; at the same time, health is no longer simply the domain of conventional medicine, nor the clinic.

More broadly, the social management of health itself is losing its anchorage in collective social relations and shared knowledge and practice, whether at the level of the local community or through state-funded socialized medicine. This individualization of health is both culturally driven and state sponsored, as the promotion of 'self-care' demonstrates. The very technologies that redefine health are also the means through which this individualization can occur—through 'e-health', diagnostic tests, and the commodification of restorative tissue, such as stem cells and cloned embryos.

This Series explores these processes *within* and *beyond* the conventional domain of 'the clinic', and asks whether they amount to a qualitative shift in the social ordering and value of medicine and health. Locating technical developments in wider socio-economic and political processes, each text discusses and critiques recent developments within health technologies in specific areas, drawing on a range of analyses provided by the social sciences. Some will have a more theoretical, others a more applied focus, interrogating and contributing towards health policy. All will draw on recent research conducted by the author(s).

The Health, Technology and Society series also looks towards the medium term in anticipating the likely configurations of health in advanced industrial societies and does so comparatively, through exploring the globalization and the internationalization of health, health inequalities and their expression through existing and new social divisions.

This book makes a valuable contribution to the Series through its focus on the gendered dynamics of the use of information and communication technologies (ICTs) in health care. Contributors to the book address two key questions. First, how is our understanding of gender informed by analysing the socio-technical relations of ICTs in health care? Second, how does an appreciation of gender relations improve analysis of the ways in which ICTs are developed, implemented, and used in health care contexts? This book brings together insights from three different fields: gender and technology, technology and health, and gender and health; and applies them to a variety of formal and informal work settings. The book also has a strong international focus that provides for a rich, comparative analysis that will appeal to many different academic and policy communities working across the gender/ICT/health care boundaries.

Andrew Webster and Sally Wyatt

Acknowledgements

This book grew out of a strong interest in gender issues shared by many ACTION for Health team members. ACTION for Health was a four-year, $3 million research project funded by the Social Sciences and Humanities Research Council of Canada (SSHRC), through the Council's Initiative for a New Economy Collaborative Research Initiative funding programme. Research undertaken through ACTION for Health addressed the role of technology in the production, consumption, and use of health information, and considered issues from both a policy and practice perspective. In contrast to many other academic funding sources, the funding programme that supported our collaboration and research, which formed the basis for all but two chapters in this collection (which were supported through grants from the UK's Economic and Social Research Council), stressed engagement with community partners. Contributors to this book worked with an enormous range of community partners, from large agencies such as hospitals to smaller agencies such as clinics and community centres. Hence, our thanks go first and foremost to all of our community partners, without whose cooperation and enthusiasm, our research would not have been possible. We would also like to thank our funders for recognizing the value of engaging varied types of partners and agencies in research.

ACTION For Health was an interdisciplinary project, which included not just a range of agencies, but also academics from varied disciplines. As our work on this book progressed, we were often delighted on the one hand by the diversity of perspectives represented in the contributions, and on the other hand, challenged by that same diversity. Addressing those challenges required ongoing contact over time, and we are grateful to SSHRC for providing a funding vehicle that allowed our team to meet several times over the life of this project to discuss contributions and allowed the editors to meet to discuss and draft various aspects of the manuscript.

Great projects usually revolve around great team members and staff, and we have had the pleasure of working with both. It has been a pleasure to work with the contributors to this book, and our interactions with the contributors have been supported over the life of the project by Alison Robb, who helped organize some of our early meetings; Melanie Klingbeil, who worked with us during the early stages of the project; and Lindsay Lynch, who has assisted us in varied ways over the life of the project, which have included communicating with contributors, undertaking research required for completion of the book proposal, providing a critical read of pieces at various stages, and editing, formatting, and compiling the final manuscript. Without such talented team members, our work would have been much more difficult, and we are grateful to both our contributors and staff members for their contributions over time.

Our series editors, Andrew Webster and Sally Wyatt, have been enthusiastic about the project from the start, responded quickly to our queries and have had great suggestions, and, together with Palgrave staff members Philippa Grand and Olivia Middleton, have been a pleasure to work with. We are grateful to have had the support and intellectual space to engage in the questions this project has allowed us to ask, and look forward to future discussions and debates we hope it will contribute to.

The editors and publisher gratefully acknowledge the Public Health Information Development Unit, University of Adelaide http://www.publichealth.gov.au/, for granting permission to adapt and reproduce Figure 2.1, 'Remoteness Scale, Australia'.

List of Contributors

Hugh Armstrong, PhD, is Full Professor cross-appointed to the School of Social Work and to the Institute of Political Economy at Carleton University in Ottawa, Canada. He publishes widely on the political economy of health and health care and on women and work. His recent articles have appeared in journals such as *Sociology of Health and Illness*, *Sociological Review*, *International Journal of Canadian Studies*, *Canadian Journal of Law and Society*, and *Nursing Inquiry*. His latest books, written with Pat Armstrong, include *Wasting Away: The Undermining of Canadian Health Care*, published by Oxford University Press, and *Critical to Care: The Invisible Women in Health Services*, published by the University of Toronto Press and co-authored by Krista Scatt-Dixon. Address: School of Social Work, Carleton University, Ottawa, Canada K1S 5B6; email: hugh_armstrong@carleton.ca.

Pat Armstrong is Professor of Sociology and Women's Studies at York University in Toronto, Canada, and holds a Canadian Health Services Research Foundation Chair in Health Services. She is author, co-author or editor of more than 15 books on women's work, health care, and social policy, most recently *Critical to Care: Women's Invisible Work in Health Services* (University of Toronto Press, 2008). She has qualified as an expert in women's work and in pay equity in a dozen cases brought before tribunals and courts. She has served as Chair of the Department of Sociology at York University and Director of the School of Canadian Studies at Carleton University. She currently chairs the Women and Health Care Reform Group, funded by Health Canada and is Acting Director of a Centre of Excellence for Women's Health.

Ellen Balka is a professor in Simon Fraser University's School of Communication, where she also serves as Director of the Assessment of Technology in Context Design Lab. She is Senior Research Scientist at Vancouver Coastal Health Research Institute's Centre for Clinical Epidemiology and Evaluation, and was recently awarded a Michael Smith Foundation for Health Research Senior Scholar award. Ellen's research focuses on issues related to the use of information technology for the production, consumption and use of health information. She is particularly interested in issues related to gender and inequality, and whether or not information technology contributes to or challenges gender and ethnic inequalities. She served as the principal investigator of the ACTION for Heath research programme, a four-year $3 million project funded by Canada's Social Sciences and Humanities Research Council of Canada, between 2003 and 2008; email: ellenb@sfu.ca.

Leslie Bella is a research professor at Memorial University of Newfoundland, where she taught in the School of Social Work until her retirement. She continues to research issues related to populations marginalized by class, gender, race, and geography, and has a continued interest in professionalized health occupations.

Eileen Green is Professor of Sociology and Director of the Centre for Social and Policy Research at the University of Teesside's Social Futures Institute. Her research interests focus around issues of social in/exclusion, especially gender issues. Recent projects have addressed topics such as midlife women and innovative health technologies; and the impact of social exclusion on women's labour market participation. She is currently working on Digital Villages, a community partnership project funded by the UK Big Lottery Fund. Co-edited books include: *Gendered by Design? Information Technology and Office Systems* (Taylor and Francis, 1993, with Jenny Owen and Den Pain), *Virtual Gender: Technology, Consumption and Identity* (Routledge, 2001 with Alison Adam) and *Youth, Risk and Leisure: Constructing Identities in Everyday Life* (Palgrave Macmillan, 2004 with Wendy Mitchell and Robin Bunton) Address: Centre for Social and Policy Research, School of Social Sciences and Law, University of Teesside, Borough Road, Middlesbrough, UK, TS1 3BA; email: E.E.Green@tees.ac.uk.

Frances Griffiths was trained in medicine at the University of Cambridge and Kings College Hospital, London, and went on to become a general practitioner in Stockton-on-Tees. While working as a GP she undertook her PhD at the University of Durham, Department of Sociology and Social Policy. She joined the University of Warwick in 1998 developing her research interest in the impact of technology on perceptions of health while directing a primary care research network. In 2003, she was awarded a UK Department of Health National Career Scientist Award to develop a programme of research on Complexity and Health that builds on her interdisciplinary research experience. Address: Health Sciences Research Institute (HSRI), Warwick Medical School, University of Warwick, Coventry, CV4 7AL, UK; email: f.e.griffiths@warwick.ac.uk.

Michelle Hall is a researcher in the Faculty of Business at QUT, who joined the Service Leadership and Innovation Research Program to be involved in this project. Her Masters research focuses on the relationship between consumption practices and place-based communities, exploring ways in which servicescapes can help to build social networks. Address: Faculty of Business, Queensland University of Technology, GPO Box 2434, Brisbane, 4001; email: ml.hall@qut.edu.au.

Roma Harris is Professor in the Faculty of Information and Media Studies at The University of Western Ontario. She has written about the impact of technological change on women's work in libraries and has been involved in a number of studies of help information-seeking by abused women. Currently, her work focuses on health help-seeking in rural communities and she is leading the 'Rural HIV/AIDS Information Networks Project' funded by the Canadian Institutes of Health Research. Harris is a founding member of two agencies to serve abused women as well as London's Centre for Education & Research on Violence Against Women and Children. Email: harris@uwo.ca.

Flis Henwood is Professor of Social Informatics in the School of Computing, Mathematical and Information Sciences at the University of Brighton, where she heads the Social Informatics Research Unit (www.brighton.ac.uk/cmis/research/groups/siru). Her most recent work focuses on the implementation and use of information and communication technologies in health care. She has published widely on e-health issues in both academic journals and peer-reviewed professional/practitioner journals. She co-edited (with Ellen Balka) an 'e-Health' special issue of the journal *Information, Communication and Society* in 2005. Her publications have appeared in a range of disciplinary fields, including sociology of medicine, social policy, information systems, and media studies. Email: f.henwood@bton.ac.uk.

Gael Le Jeune is a statistical analyst at Statistics Canada, where she assists researchers who are working on micro-data from large-scale longitudinal surveys on social matters. She has studied Economics and Demography, and her own research focuses on every aspect of women's work. She is especially interested in the ways to capture women's work, the invisibility of which has important consequences on gender relations and women's health. She has contributed to research on female migration in West Africa and also on the 'data gap' in women's occupational health in Canada.

Susan Leggett is a research assistant in the Faculty of Business at Queensland University of Technology. Since 1999, she has participated in a series of projects investigating issues affecting the lives of rural and remote Queenslanders, and the potential for technologies to alleviate some of the barriers caused by geographical isolation. Address: Faculty of Business, Queensland University of Technology, GPO Box 2434, Brisbane 4001, Australia; email: s.leggett@qut.edu.au.

Antje Lindenmeyer has an interdisciplinary background in gender, social, and literary studies, with a special interest in autobiographical narratives, and holds a PhD in Women and Gender Studies from the University of Warwick. She is now a research fellow in primary care at Warwick Medical School and has contributed to the areas of patient involvement/empowerment, the self-management of chronic conditions and coping with disfigurement and other appearance issues. Her main research interest is in family narratives of health and illness and their impact on health care encounters. Address: Health Sciences Research Institute, Warwick Medical School, Coventry, CV4 7AL, UK; email: Antje.Lindenmeyer@warwick.ac.uk.

Karen Messing, Ph.D. is retired Professor of Ergonomics at the Université du Québec in Montréal, Canada. Her current research focuses on applications of gender-sensitive analysis in occupational health and constraints and demands of work in the service sector. Dr Messing co-directs a research partnership with three Québec unions oriented towards improvement of women's occupational health. Ongoing studies include effects of mobility among workers exposed to prolonged standing, musculoskeletal problems in women's non-traditional jobs, and ergonomic exposures of cleaners. She is the author of over 100 peer-reviewed articles and of *One-eyed Science: Occupational Health and Working Women*, of *Women's Health at Work* (with Kilbom and Thorbjornsson), of the WHO's *Gender Equality, Work and Health: A Review of the Evidence* (with Östlin) and editor of *Integrating Gender in Ergonomic Analysis* (published in six languages). She presides over the Gender and Work Technical Committee of the International Ergonomics Association.

Zena Sharman is a PhD Candidate in the Interdisciplinary Studies programme at the University of British Columbia. Her research focuses on the working lives and working conditions of women workers in the health care system, which she analyses from a critical feminist perspective. Her recent work examines the experiences of home support workers in rural and remote communities. She holds doctoral awards from the Canadian Institutes of Health Research and the Michael Smith Foundation for Health Research. Address: Department of Health Care and Epidemiology, Faculty of Medicine, University of British Columbia, 5804 Fairview Avenue, Vancouver, BC V6T 1Z3; email: zsharman@interchange.ubc.ca.

Lyn Simpson is Assistant Dean in the Faculty of Business at Queensland University of Technology, Brisbane, Australia. Lyn has conducted extensive research and consulting projects in the areas of social and policy implications of communication technologies for sustainable development in rural communities. Lyn is particularly interested in the impacts of communication technology on community social capital, capacity building, and rural community development. Her most recent projects have focused on the ways that ICTs can support health literacy and health outcomes particularly in remote and indigenous communities. Address: Faculty of Business, Queensland University of Technology, GPO Box 2434, Brisbane, 4001; email: le.simpson@qut.edu.au.

Sally Wyatt is Professor of Digital Cultures in Development at Maastricht University and a senior research fellow with the Virtual Knowledge Studio for the Humanities and Social Sciences, Royal Netherlands Academy of Arts and Sciences. Her research focuses on the relationship between technological and social change, focusing particularly on issues of social exclusion and inequality. She has been president of the European Association for the Study of Science and Technology (2000–2004). Together with Nadine Wathen and Roma Harris, she has edited *Mediating Health Information: The Go-Betweens in a Changing Socio-Technical Landscape* (Palgrave Macmillan, 2008). Address: Cruquiusweg 31, 1019 AT Amsterdam; email: sally.wyatt@vks.knaw.nl.

Introduction: Informing Gender? Health and ICTs in Context

Flis Henwood, Eileen Green, and Ellen Balka

The growth of information and communications technologies in health care

Throughout the world, information and communication technologies (ICTs) are increasingly being used in health care delivery. Information systems to support a wide range of health care activities including the collection of health information, the distribution of health information to citizens, and improved diagnoses and care have been developed to reflect the new political priorities of increased accountability, patient and public involvement, and evidence-based medicine.

As the computerization of health care occurs, new social and political issues and challenges arise. The introduction of information systems into health sector workplaces may change the way that work is undertaken. Increased availability of online health information may change doctor–patient relationships and raise a plethora of jurisdictional and legal issues. New imaging technologies such as magnetic resonance imaging (MRIs) may reveal previously unseen parts of bodies, giving rise to new questions for both practitioner and patient alike. As new information systems are developed, important decisions are also being made about what data they should be used to collect—decisions that will influence what we are able to 'know' about the health of populations for many years to come.

A key area of social change associated with new health information systems is the changing boundaries and relationships between traditional 'producers' and 'users' of health information. Thus, health information 'consumers' today might include health care providers (such as doctors, nurses, physiotherapists, social workers, etc.) who increasingly access online support and advice for their practice, as well as the more traditional consumers including patients, family members, and carers. Similarly, health information 'producers' no longer only comprise health providers and specialist health information providers but are a diverse group that may include researchers, administrators, patients, carers, and patient organizations.

Health information is now produced from a wide variety of sources. Health practitioners may produce health information by taking a patient's vital signs and entering those into a computer system. Researchers and administrators become health information producers when they conduct analyses of aggregated health information for research or administrative purposes. Patients, their friends, and/or family members may become health information producers when they 'blog' (produce an online weblog) in a way that describes the progression of a disease, or through their contributions to other forms of online discussion or information dissemination. Organizations (such as community groups, disease-based agencies, doctors' offices, or hospitals) may become health information producers by putting health information online via web pages that may contain anything from x-rays or other images to textual information about health.

As boundaries and relationships between producers and consumers of health information become ever more complex, interesting new work about health information 'intermediaries' (Wathen, Wyatt, and Harris, 2008) points to the ways in which both human (professional and lay) and technological intermediaries are becoming increasingly important in mediating the relationship between producers and consumers of health information.

In addition to the changes we are seeing in health information producer–consumer relationships, the computerization of the health sector is also associated with an increased range of sites of health care delivery. As new information systems become networked, and as opportunities for distributed care increase, health care is increasingly accessed away from the central hospital—in community clinics and even at home. As the diversity of locations in which health care is delivered increases, there are implications for both the range and the type of people who become enrolled in the delivery of health care, and for their involvement in producing and consuming health information to support that care.

Why gender?

There is now a substantive body of work that addresses the social and organizational contexts within which new ICTs are produced and used in health care. Articles can be found in journals that span health informatics, medical sociology, and media and communications, as well as journals focused specifically on social aspects of ICTs.[1] In addition, collections that focus either on the use of information systems in hospital settings (Berg, 2004) or on the use of the Internet and other digital platforms by health providers and/or consumers across a range of settings (Murero and Rice, 2006; Nicholas, Huntington, Jamali, and Williams, 2007) are increasingly common. Essays focusing specifically on the role of ICTs in consumer or patient 'empowerment' can be found in a collection on new technologies in health care by Webster (2006) and in a special issue of the journal *Information, Communication & Society* on e-Health (Henwood and Balka, 2005). However, despite this burgeoning of literature on social aspects of ICTs in health care, there has, as yet, been no systematic collection of essays that address the *gendered* dynamics of these socio-technical changes. This book breaks new ground, therefore, in addressing how gender mediates the socio-technical relations surrounding the use of ICTs in health care. As the title to this introductory chapter suggests, the book explores how the cases addressed here can be helpful in 'informing gender'. This title speaks to the two key questions that chapters in the book seek to address: how can our understandings of gender be informed by exploring the socio-technical relations of ICTs in health care, and, conversely, how far can an appreciation of the ways in which gender works, inform, and improve our understanding of how ICTs are being developed, implemented, and used in health care contexts? We will return to these questions in the final, concluding chapter of the book, where we suggest ways forward for research in the field. Here, we wish to identify the different bodies of literature that have been influential in framing these questions and discuss how they shape the approaches taken in the various chapters that make up this collection.

The three key bodies of literature that have helped shape our approach to this book can be broadly described as the 'gender and technology', the 'technology and health', and the 'gender and health' literatures. We do not intend to provide an exhaustive review of each of these literatures in this introduction but rather to show how each has contributed to the framing of the book, which seeks to explore, perhaps for the first time in one collection, the intersection of gender, technology, and health.

Gender and technology

There is a significant body of work exploring the gendered relations of technology, including ICTs (notable contributions include those by

Wajcman, 1991, 2004; Grint and Gill, 1995; Lie, 2003), but very little that explores the gendered relations of ICTs in health care settings, specifically. Here, we provide a brief overview of some of the key themes from literature about gender and technology that have been instructive in developing our understandings of the social dynamics involved in the use of ICTs in health care settings.

Early scholarship about gender and technology suggested that women and men are differently affected by technological change, and research undertaken in the 1980s helped challenge the view that technologies are neutral and value free. In line with the 'social shaping' approach to understanding technology–society relationships (MacKenzie and Wajcman, 1984), technologies were understood as shaped to reflect and reinforce hierarchical relationships—of class and gender, in particular (Cockburn, 1983, 1985; McNeil, 1987; Hacker, 1990; Wajcman, 1991). About the same time, both social scientists and computer scientists in the OECD countries began considering what this insight might mean for both work practices and system design, and new collections exploring the gender and design relationships were published (Olerup, Schneider, and Monod, 1985; Green, Owen, and Pain, 1993; Adam, Emms, Green, and Owen, 1994), whilst calls for more women in technology and design were commonplace.

Throughout the 1990s, work on gender and technology increased, with more nuanced accounts of 'the gender–technology relation' (Grint and Gill, 1995) being put forward. This body of work emphasized the cultural as well as the material relationships between gender and technology and suggested that, precisely because the equation between masculinity and technology was symbolically maintained, it may not be overcome simply by the presence of more women *in technology* (Henwood, 1996; Faulkner, 2001). Also in this decade, the rise of the Internet offered opportunities for these more cultural analyses of the gender–technology relation to flourish, and, as Wajcman (2004) has argued, the decade marked the emergence of what came to be referred to as 'cyberfeminism', as Haraway's 'cyborg' metaphor (Haraway, 1985) became an increasingly popular tool in postmodern feminist analyses of technology. Whilst some notable feminist texts in this period saw the Internet and other ICTs as unambiguously empowering for women (see, for example, Plant, 1997), several significant pieces of work continued to interrogate claims for the emancipatory potential of the Internet for women (Balka, 1997; Kirkup, Janes, Woodward, and Hovenden, 2000; Green and Adam, 2001). Either way, by the end of the decade, the move towards constructivist understandings of the gender–technology relation was almost universal, not least because it enabled recognition of women's agency in relation to technology.

This constructivist turn in feminist technology studies was in line with developments elsewhere in technology studies (e.g., Bijker, Hughes, and Pinch, 1987) and gave rise to understandings of the gender–technology relation that relied on neither technological nor social determinism, but rather stressed the mutually constitutive and always contingent nature of the relationship. Instead of focusing on how existing gender relations are simply reflected in technologies, feminists began to understand technology as both a cause and a consequence of gender relations. As Wajcman (2007: 293) has argued, 'gender relations can be thought of as materialized in technology, and gendered identities and discourses as produced simultaneously with technologies'. In this collection, we seek to analyse the ways in which gender and technologies are co-produced in a range of different health care settings and practices.

Technology and health

Several bodies of related literature, including medical sociology and medical anthropology; science and technology studies (STS); computer supported cooperative work (CSCW); medical informatics; and information systems, have contributed to an emergent area of scholarship known as socio-technical perspectives on health informatics. This area is characterized by several shared assumptions and practices that have also shaped much of the work in this collection. First, there is a shared assumption that one of the key reasons why new technologies do not automatically result in the social and organizational improvements claimed for them is because of the lack of 'fit' between the technologies and the use contexts. It follows from this that, central to the successful adoption or implementation of new technologies is a detailed understanding of users and use contexts. Just as Bijker et al. have argued that machines only 'work' because they have been accepted by relevant social groups (Bijker et al., 1987), and Latour has stated that 'the fate of...machines is in later users' hands' (Latour, 1992), Berg, with reference to health care, has argued that 'many of the computer-based PCISs (Patient Care Information Systems) that litter the cemetery of "failed attempts" were *bound* to fail because the model of health care work inscribed in these tools clashed too much with the actual nature of health work' (Berg, 2004: 48).

This lack of fit between technologies and work practice has become a central theme in the socio-technical approach to health informatics literature. Research has focused on questions of user acceptance of, or resistance to, new technologies, and/or to understanding the work involved in making the technologies fit the work of specific user groups (Jones, Henwood, Gart, and Gerhardt, 2003; Timmons, 2003a). Much of this work draws upon and applies concepts developed outside of health informatics. For example, notions of user 'appropriation' (Suchman and Jordan, 1989) and 'domestication' (Silverstone and Hirsch, 1992) of technologies were first developed within STS and media studies (respectively) and have been widely used to draw attention to the ways in which users 'make technology their own' (Lie and Sorensen, 1996). The concept of 'articulation work' (Fujimura, 1987; Star, 1991) has also been used to highlight the work needed to make technologies fit for purpose, and the linked concept of 'invisible work' (Star, 1991; Star and Strauss, 1999) has been used to explain why articulation work is so often overlooked in the design and development of new technologies.

A further shared assumption in much of this socio-technical literature in health informatics follows from these understandings and has implications for design practice. As users' roles in shaping technologies to fit their needs have been increasingly recognized, arguments for, and examples of, 'design in use', 'co-design', and 'co-production' have been promoted to capture a practice that recognizes the artificial separation between design and use (Hartswood et al., 2003). It is useful to reflect on the words of Suchman (2002) to understand the linkages between the concepts of appropriation, articulation work, and design-in-use:

> if technologies are to be made useful, practitioners of other forms of work must effectively take up the work of design, as those activities currently glossed under the notion of technology adoption; that is, appropriating the technology so as to incorporate it into an existing material environment and set of practices. Integration, local configuration, customization, maintenance and redesign on this view represent not discrete phases in some 'system life cycle' but complex, densely structured courses of articulation work without clearly distinguishable boundaries between.
>
> (Suchman, 2002: 93)

Although there is some interesting feminist work that links notions of articulation work and invisible work to gendered work practices (Star, 1991; Wagner, 1993), and some work that has begun to explore how gender is articulated in user resistance to new technologies in health care (Henwood and Hart, 2003), there has been little systematic exploration of the gendered dynamics involved in design and/or use of ICTs in health care. The contributions in this collection take these explorations further than has been achieved before in an attempt to develop new insights and suggest some new areas for research (see the concluding chapter).

Another significant body of work that has helped shape some of the contributions in this collection also draws on work in STS (see, e.g., Star and Ruhleder, 1996) to assess the significance of large-scale information infrastructures for structuring work in such a way as to render invisible the social character of scientific knowledge production. In a health care context, large-scale infrastructures such as hospital information systems typically have numerous classification systems built into them. For example, a hospital information system may use the International Classification of Diseases (ICD) 9 or ICD 10 to track the diagnoses of patients, which in turn may be used as part of a billing process in private health care systems. Changing classifications systems from ICD 9 to ICD 10 has implications for what data are collected about what illnesses, which in turn has implications for how one understands health problems. Such classification systems also reflect social values and norms. For example, until 1973, homosexuality was listed as a psychiatric disorder in the American Psychiatric Association's *Diagnostic and Statistical Manual of Mental Disorders*,[2] but was removed from the DSM as values and beliefs about homosexuality changed (Spitzer, 1981). Work in STS has shown how classification systems and other large-scale infrastructural technologies have consequences: they carry economic and political weight, and yet they often enter into common discourse with an air of neutrality that obfuscates their essentially value-laden and political nature. Several contributions to this collection draw on these insights to explore how gender is (re)produced when such systems produce ostensibly neutral scientific data on health status disease probabilities or even the status of the health care system itself.

Gender and health

Another influential body of literature, in terms of the framing of this book, is focused around the area of gender and health. Here, there are several well-rehearsed themes that are relevant to our concern with the three-way intersection of gender, health, and technology. A major theme concerns the way in which gendered divisions of labour (in both paid and unpaid work) and gender relations, more generally, result in gendered health conditions and practices. For example, women experience the health consequences of domestic violence and rape have a higher level of reported anxiety and depression than men, and often experience poor attitudes from health professionals in the medical encounter (Doyal, 2001). In addition, because women take responsibility for the health of others (including both men and children), they often end up ignoring their own well-being or delaying necessary health treatments. For example, Guillemin (2004) has shown that women's symptoms of heart disease are often overlooked because of its representation as a male disease, and women often delay seeking a diagnosis because they are too busy looking after others. Men's health is also gendered, of course. For example, a growing body of literature on men's health has drawn attention to the links between working-class male occupations and the specific health risks men face (Doyal, 2001) and between 'risky behaviours' and the performance of hegemonic masculine identities (Canaan, 1996; Schofield et al., 2000), and to men's reluctance to ask for help or support from the health services (Banks, 2001; Lloyd, 2001; Biddulph and Blake, 2001; see also Henwood and Wyatt, this volume). What part does technology play in these gendered practices, and how might gendered norms and values be shaping the ways in which new technologies are used in these areas of health care practice? These questions are taken up within this collection.

Another theme in the gender and health literature that is relevant for our framing of this book concerns gendered divisions of labour in health care work, specifically. Here, key issues include the gendered division between medical and nursing work (often referred to as the difference between 'curing' and 'caring'), as well as the invisibility of women's unpaid care work (Graham, 1983; Grant et al., 2004). How are the implementation and use of new ICTs in both formal and informal care work impacting on such gendered divisions of labour, and how does gender shape the ability of different groups of carers to influence the technological design and implementation process? In the context of health sector restructuring, where not only are new technologies being applied to formal medical and nursing processes, but also these same technologies are supporting the movement of health care delivery to the informal care sector, such questions are both critical and timely and are explored in various ways in different parts of this collection.

A third area of the gender and health literature that has been influential for us in framing this collection concerns 'the gendered body'. Gender and health scholars have shown how the body is gendered through health care practices such as screening (Howson, 1998), often linking debates about screening to the wider literature on risk and surveillance, showing how women seek to manage risk in a context where the 'risky body' is typically represented as a feminine body. These ideas are addressed in this collection where, for example, Green, Griffiths, and Lindenmeyer (Chapter 9) make the role of what they term 'visual technologies' in the production of the gendered body more explicit.

Reflections on theoretical approaches

As an edited collection, this book does not adopt one theoretical position or approach to analysing the relationship between gender, health, and technology. Indeed, each of these three concepts is treated in diverse ways, reflecting, on the one hand, the multidisciplinary and interdisciplinary nature of its contributions, and, on the other, the ongoing debate in the social sciences about the relative significance of structure and agency in the constitution of social relations and identities.

Reflecting the eclecticism of gender studies more generally, contributors take a range of disciplinary and theoretical approaches to the study of the gendered dynamics of both ICTs and health. For example, some contributions emphasize structural factors such as gendered divisions in both formal and informal health care work (see, e.g., Armstrong, Armstrong, and Messing, Chapter 7). Here, issues explored include how the implementation and use of ICTs affects where and how the boundary between formal and informal care work is drawn, or how the distribution of such work between women and men is made. Other contributors are more concerned with the performance of gendered identities and other forms of agency, especially in relation to the consumption of health information and ICTs. Examples here include explorations of how gendered identities are articulated in discourses surrounding the uptake and use of the Internet for health information seeking (see Henwood and Wyatt, Chapter 1) and in the negotiations that occur when health care workers try to achieve a 'fit' between a new technology and their preferred work practice (see Sharman, Chapter 5).

Health, too, is treated in diverse ways in this collection. Some contributions (see Armstrong et al., Chapter 7) are firmly rooted in a 'social determinants of health' perspective, and others emphasize a more constructivist approach to health, social care, and well-being (see Henwood and Wyatt, Chapter 1; Simpson et al., Chapter 2). More often, however, contributors are keen to try and understand how experiences of health that may appear as individual are actually shaped in the context of wider social structures and discourses. This is particularly important in the context of current government health policies that tend to stress individual responsibility for the maintenance of health and well-being whilst downplaying the wider social and environmental factors that contribute to ill-health.

Treatment of technology and ICTs by contributors is similarly diverse. Some contributors are concerned with identifying the impacts of technologies on particular areas of health care, for example, work practice (see Armstrong et al., Chapter 7; Sharman, Chapter 5 and Balka, Chapter 6), and others are more interested in exploring how technologies are themselves constituted in specific health contexts (Green et al., Chapter 9; Henwood and Wyatt, Chapter 1).

Thus, there is no single theoretical position that is promoted in this collection. Whilst some contributors favour more determinist and others more constructivist approaches, more commonly, we find an engagement in a process of analysis that seeks to strike a balance between these two extreme positions, showing where and how spaces are opened up for productive and empowering reconfigurations of the gender, technology, and health relationship. In the concluding chapter, we reflect on what such analyses might mean for debates in the field, for policy and practice and for future research.

The book's history and structure

The idea for this book emerged as a result of authors working together across disciplinary boundaries and continents through the ACTION for Health project, an international research project funded by the Social Sciences and Humanities Research Council of Canada (SSHRC) between 2003 and 2008.[3] The project explored the use of ICTs as a means of information delivery in a variety of formal and informal health care contexts, allowing analysis of the 'situated' uses of ICTs in health care. The ACTION for Health project provided opportunities for the authors whose work is represented here, to see how the treatment of the three key concepts addressed in the book—health, technology, and gender— are understood differently both within and across academic disciplines and fields of practice. ACTION for Health researchers spanned the disciplines of information science, communication studies, health studies, political science, sociology, biology, medicine, psychology, computer science, law, and more, with a core group coming from STS. Also, key to ACTION for Health research was working in partnership with practitioners—in either health and/or information fields. This provided a further interesting challenge as academic concepts were tested in practice, and practice-based problems helped shape research agendas. Empirical research, which formed the basis of chapters in this collection, was undertaken in Canada, the United Kingdom, and Australia, in rural and urban settings, and in a range of formal and informal health care and/or health information settings, including hospitals, community clinics, community centres, libraries, and homes.

Chapters in this collection cover three broad themes—health information seeking, informal contexts and gendered care giving; formal health care settings, ICTs and gendered paid work; and ICTs and gendered ways of knowing about health. The first four chapters (Henwood and Wyatt, Simpson et al., Bella, and Harris) all consider how gender and ICTs are intertwined in the context of informal care giving. A key theme here is how technology is used in the process of health information seeking in non-clinical, informal health care settings. Particular emphasis is placed within these chapters on how gender mediates the use of the Internet as a medium for accessing health information, on the emergence of the 'informed patient', and on the role of information intermediaries. This book begins with a chapter by Flis Henwood and Sally Wyatt, who explore how gender is implicated in the different (and varied) positions individuals take up in relation to the informed patient (IP) discourse. They argue that the IP discourse can be understood as destabilizing the traditional gender binary—where health and health management is traditionally figured as feminine and computer technology as masculine. However, such destabilization does not necessarily lead to transformations in gender relations and identities. Instead, it can give rise to a set of gender re-inscription processes, which acts to inhibit the very same transformative practices suggested by the gender–health intersection within the IP discourse. Henwood and Wyatt present examples of such processes occurring in different sites and illustrate how gender is inscribed in relation to people, technologies, and health care practices in ways that are always deeply contextual, negotiated, and relational.

Lyn Simpson, Michelle Hall, and Susan Leggett also take up the theme of how gender mediates the uptake and use of the Internet—in their case, by health intermediaries in rural and remote Queensland. The authors start by discussing the importance of social support and social networks for maintaining health and well-being, and how activating and enhancing these informal networks can assist in empowering individual and communities alike. They argue that, although this understanding is well-established in mental and public health research, government-led health programmes often still overlook the importance of social support in health care and focus, instead, on information provision alone as a means to encourage preventative health behaviours and self-care. Such a focus is, they argue, encouraged by the presumed efficiencies of ICTs, such as the Internet, that are seen as cost-effective means of providing health information to a large number of consumers. In this chapter, Simpson, Hall, and Leggett draw on research into the work of health information intermediaries in rural and remote Queensland, Australia, to argue that assumptions about gender may account for this apparent indifference to the role of social support and informal helping in health care. Thus, whilst all health intermediaries (male and female) were found to demonstrate a diverse range of skills as required by the specific context, rather than as determined by their individual gender identity, assumptions about gender roles continue both to obscure and to diminish the importance of this flexibility in ways that result in a lack of access to appropriate technologies and training to meet these workers' needs.

Leslie Bella also addresses the complex gender dynamics involved in care work, in this case, in a community centre (in Newfoundland, Canada) where community residents are learning to use public access computers. Like Henwood and Wyatt, and Simpson, Hall, and Leggett, Leslie Bella describes and analyses the workings of gender in a way that goes beyond traditional binary conceptualizations. Bella explores the gender–health–technology relation by addressing the significance of (and limitations to) an 'ethic of care' in supporting the use of public access computers at a site in a low-income neighbourhood. This idea of an ethic of care provides fruitful guidance for conceptualizing the gender–health relation. Building on the work of Joan Tronto (1993), Bella speculates that, in relation to technology, one might expect 'executive caring' to be men's work and 'caring about' and 'being cared for' to be women's work. However, the relation between this ethic of care and gender proves to be more nuanced (and more interesting) in the context of the vulnerable population of this low-income neighbourhood. Bella discusses how both men and women begin their interaction with the community access computers by 'being cared for', and both also move to a point where they can and will 'care for others'. Finally, responsibility for executive caring is shared between the centre's committees (involving both men and women) and the Community Access Program intern (who is sometimes male and sometimes female).

Roma Harris' chapter consolidates the exploration of the role of health information intermediaries developed within the collection—this time exploring how the unpaid or 'caring' work of women is often exploited to support the effective use of the Internet in this regard. Drawing on a series of empirical studies undertaken over five years in Ontario, Canada, Harris explores gender relations in the care work performed by formal and informal health information intermediaries, work that heavily involves the Internet. Harris explores questions about how gender roles are transformed or re-inscribed as a result of ICT use in health care work. She discusses how gender is mobilized in the organization and delivery of ICT-based health information and services, specifically in the context of library services. Although the mandate of libraries is to retrieve, organize, and provide their users with access to information, they are in many respects seen only as 'bit players' in the larger health information landscape. This is because, as institutions, libraries are sometimes perceived to be anachronisms, no longer necessary because everything one could want to know is available through the Internet. Furthermore, because librarians (the majority of whom are women) are perceived to be practitioners of a 'not technology' occupation, they are not seen to have much to offer design and support of ICT-based health information services. Heavy investments are made in the creation of technology-supported e-health infrastructures, even though there is growing evidence to suggest that they do not necessarily replace the need for personal intervention and intermediation on the part of many citizens who are in search of health information. The analysis in this chapter gives the reader insight into how conceptions of the gender–technology relation is a factor underlying the assumptions of decision makers, and in turn, how these assumptions—which contradict reality, especially given women's important role as health information intermediaries—work to reinforce gendered ways of looking at the world.

Moving into more formal health care contexts, chapters by Sharman and by Balka explore, in their different ways, how gender mediates the adoption and use of ICTs in hospital and clinical settings. A key issue addressed by these authors is the gendering of organizational and professional hierarchies and the influence of these patterns on staff and users' abilities to influence and shape ICT implementation and use. Zena Sharman explores gender and ICT relations in an emergency department and focuses on the implementation of a computerized PCIS. She explores nurses' struggle to articulate how the new technology might fit with their gendered work identity. Drawing on data from a qualitative study, she discusses how, working in a context of health system restructuring, nurses define their skilled caring work in contrast to both low-status, feminized administrative work and impersonal masculinized technology. In this way, the PCIS becomes a site of tension and resistance as emergency department nurses articulate the role of technology in nursing work practice.

Ellen Balka covers some similar ground to Sharman, but in this case drawing on over five years of fieldwork observing ICT implementations in clinical settings to explore the ways that gendered workplaces and work practices shape how new technologies are implemented and used. Balka argues that health sector information technology is often an overlooked aspect of both work practices and professional practices in health care, and studying its use, by health sector workers organized within a sex-segregated labour force, brings numerous intersecting issues to light. Such technology can alter work organization, professional practice, and the movement and use of information among care teams who cooperatively manage health. Drawing on a series of vignettes from several years of fieldwork concerned with technology implementations and use in health sector workplaces, Balka argues that gendered dynamics often hinder the optimum use of technologies in the support of patient care. She draws on Harding's gender framework (Harding, 1986) to explore how gender structures, symbols, and identities are produced simultaneously with technology use in health sector workplaces in ways that prevent the most efficient use of technology. She concludes by arguing for both health researchers and practitioners to intervene in the complex relations of gender, technology, and work in the interests of both improving women's work lives and the chances of success in information technology implementation projects.

Hugh Armstrong, Pat Armstrong, and Karen Messing also explore women's work in formal health care settings but their task is much broader than this. They explore how health technologies both reflect and reinforce assumptions about women and women's work whilst, at the same time, vertical and horizontal segregation in health care work leaves women out of technology-related decision-making positions, where such assumptions might be challenged. Balancing out the strong emphasis on empirical work throughout this book, the chief goal of this chapter is to raise compelling questions about the social factors that are both taken for granted and remain invisible in the technological practices in health care work. The authors systematically uncover some of the gender-related values, assumptions, and interests that are built into the structure of information technology used in health care work. They understand ICTs in health care as reflective of the medical model of health care, with its emphasis on body parts, tasks, particular forms of evidence, and dichotomous thinking about illness and care. Built into the technologies, the authors argue, are assumptions about women, about health, about skill, about evidence, about time, about space, and about care. These assumptions shape and reinforce certain ideas and practices that often fail to reflect the interests of and the differences among women. In this way, the technologies can be understood as simultaneously reproducing existing power relations and establishing new ones, with contradictory consequences for women's work and women's health.

This idea of technologies playing an active role in constructing or 'producing' health is picked up again in the final two chapters of this collection, which, together with Armstrong et al., can be understood as contributing to the third broad theme of this collection: ICTs, and Gendered Ways of Knowing about Health. Gael Le Jeune explores how information systems and indicators used in the occupational health field contribute to the invisibility of women's health problems. Le Jeune demonstrates how the information systems in use at workers' compensation boards contribute to the invisibility of women's occupational health problems through a case study of Quebec compensation data. She argues that although the databases put in place to support the workers' compensation systems are often considered 'gender sensitive', the sex of the claimants is systematically recorded, and compensated injuries statistics are usually displayed by sex. In the process of constructing compensation injury databases, there is a tragic lack of both visibility and reflection about the gendered patterns of work and working conditions in Canada that produce very different views of worker health. The usefulness of existing compensation databases in explaining women's occupational health is further limited by recent changes that have included the introduction of new technologies, such as online claim-filing systems which, Le Jeune argues, are reducing the quality of data. Le Jeune's chapter demonstrates how gendered ideas about health are inscribed in data collection processes, which reify some views of health and illness, while rendering others invisible. Thus, the information systems (here, large-scale databases) used to capture and analyse health data can be seen as playing a critical role in gendering health information and knowledge.

Le Jeune's chapter is followed by another which explores how gender is implicated in 'ways of knowing'. Eileen Green, Frances Griffiths, and Antje Lindenmeyer explore the gendered dynamics involved in the use of information technologies (here, bone densitometry and mammography) in the context of a clinical consultation. They explore how both health professionals and patients (here, midlife women) enrol the technology in the construction of risk narratives. The authors place particular emphasis on the fact that the technologies involved are 'visual' technologies and explore the implications this has for how the knowledge produced by them is 'read' or interpreted by both patients and professionals in the context of the clinical consultation. Drawing on data from an empirical study of the use of these visual technologies, the authors argue that the ways in which such images are used in the consultation can work to obscure the uncertainty in knowledge about both breast cancer and osteoporosis, whilst the technologies themselves become what Donna Haraway (1990: 205) refers to as 'the crucial tools re-crafting our bodies'. The authors address the extent to which the gendered context of the clinical consultation, with (mostly) male consultants cast in the role of 'trusted experts' to whom midlife women turn for advice, shapes the privileged status of the scan and test results above more informal ways of knowing provided by women's family, friends, and their own embodied health experiences.

The concluding chapter reflects on the chapters described above and returns to our original questions: How can our understandings of gender be informed by exploring the socio-technical relations of ICTs in health care, and, conversely, how far can an appreciation of the ways in which gender works inform and improve our understanding of how ICTs are being developed, implemented, and used in health care contexts? The concluding chapter begins to provide some answers to these questions whilst simultaneously drawing out the ongoing contestations in terms of how best the gender, technology, and health relations may be theorized. Another key goal of the concluding chapter is to explore the practice and policy-based implications of this new work and make suggestions for future research.

Notes

1. See, for example, *Information, Communication & Society; Methods of Information in Medicine, Social Science of Medicine*, and *Sociology of Health & Illness*.
2. The manual 'attempts to classify all known psychological disorders according to symptoms', and is generally used by psychiatrists to determine diagnoses of patients, and to determine what does or does not constitute a mental disorder (www.macalester.edu/~psych/whathap/UBNRP/nightmares/Terms_Glossary.html).
3. This project was funded through the Social Sciences and Humanities Research Council of Canada's Initiative for a New Economy Collaborative Research Initiative, grant titled 'The role of technology in the production, consumption and use of health information: Implications for policy and practice' (ACTION for Health), Ellen Balka, Principal Investigator. We are grateful to SSHRC for their generous support.

1
All Change? Gender, Health and the Internet

Flis Henwood and Sally Wyatt

Statistical analyses increasingly suggest that women use the Internet at similar rates to men and may be even more prolific users in the area of health information (Fallows, 2005). Such analyses might suggest that there is, therefore, no significant gender divide when it comes to Internet use in a health information context. However, more sophisticated understandings of digital divides would be alert to the dangers of such an interpretation. For example, Kvasny (2006), following Hargiatti et al. (2004), has distinguished usefully between 'the digital divide' and 'digital inequality', arguing that whereas the former refers to disparities in the structure of access to and use of information and communication technologies (ICTs), the latter is a broader concept, reflecting the ways in which long-standing social inequalities shape beliefs and expectations regarding ICTs and their impact on life chances (Kvasny, 2006: 160). This hints at the need for much more 'situated' understandings of Internet use than statistical analyses can offer. With respect to gender, it suggests that long-standing divisions and inequalities—in health, in the home, in employment and in relation to technologies—may shape how people approach use of the Internet and integrate it into their daily lives to engage in specific tasks and achieve specific ends. This chapter seeks to explain how gender and the Internet are 'co-produced' in specific use contexts and how Internet use for health information and decision making might be understood in the context of the discourse of the 'informed patient'.

By 'informed patient discourse', we are referring to a set of assumptions about responsibility for health and health information seeking that have emerged in late modern societies. These assumptions can be found in government policy documents setting out agendas for public health and self-care and, as we have argued elsewhere (Henwood

et al., 2003), versions of the discourse can be found in some academic literature, too. This 'informed patient' can be understood as
a sort of 'ideal model' of how health care consumers should engage
with health services. 'Informed patients' (IPs) are understood as taking
increasing responsibility for their own health and those of their families
and communities by actively seeking health information and making
informed choices through engagement in shared decision making with
health professionals. The Internet is seen as playing a significant role
in supporting the emergence of informed patients, offering access to
information that can assist the processes of increased self-reliance and
self-care. Our own previous work and that of others (Henwood et al.,
2003; Kivits, 2004; Nettleton et al., 2004) has identified factors facilitating and inhibiting the emergence of the informed patient and these
have offered important insights for the debate about the Internet and
so-called 'patient empowerment'. In particular, important conceptual
distinctions have been made between the Internet and information
(Henwood et al., 2003) and the need for people's engagement with
information (more or less 'reflexive') and the Internet (more or less
'domesticated') to be analysed separately (Nettleton et al., 2004). However, there has, as yet, been no systematic analysis of how gender shapes
engagement with health information and the Internet in these detailed
empirical studies. This chapter seeks to offer such an analysis, arguing
that such engagement is mediated by the informed patient discourse.

Although the IP discourse seems to hail men and women equally, we
argue that there are good reasons to suppose that gender is implicated in
the different (and varied) subject positions individuals take up in relation to it and that gender may continue to be constituted through such
subject positions and related information and Internet practices. As we
argue in the next section, both health and the Internet can be shown to
be sites of gender construction and so, when health and the Internet are
brought together in the way in which they are in the informed patient
discourse, we might expect to observe some disturbances to existing
gendered subject positions and practices. This theoretically informed
section is followed by a narrative analysis of the relationship between
gender, health and the Internet, drawing on data from an empirical
study of Internet use in the context of health information seeking and
decision making.

Health and the Internet as sites of gender construction

Health and gender can be understood as inextricably linked or even 'co-
constructed'. As argued in the Introduction (Balka, Green and Henwood,

this book), women's health is negatively affected by gendered divisions in work (both paid and unpaid), women have lower levels of income and wealth and report higher levels of anxiety and depression than men, women take the major responsibility for the health of others (including men and children) whilst ignoring their own well-being (Doyal, 2001; Daykin and Naidoo, 1995) in line with traditional notions of femininity. Men, too, experience negative consequences of socially constructed gender differences. Once again, some seem to result from what might be thought of 'gender structures'—for example, working-class men experiencing the health risks associated with working-class male occupations—whereas other differences may be thought of in relation to the formation of masculine identities. Doyal, for example, has argued that 'the development and maintenance of heterosexual male identity usually requires the taking of risks that are seriously hazardous to health' (Doyal, 2001: 1062). Other commentators explain the fact that men are more likely than women to engage in 'risky behaviours', including, the practice of unsafe sex, as being associated with the development of a hegemonic masculine identity (Canaan, 1996; Schofield et al., 2000).

It has also been suggested that the need to demonstrate a 'hard' masculinity may explain men's unwillingness both to consult a doctor when problems arise and to take health education messages seriously (Doyal, 2001: 1062; Griffiths, 1996; NHS Executive, 1998). Men are less likely than women to seek advice from peers, magazines, books and the television. Middle-class men are more likely than their working-class peers to access and respond to health promotion information from leaflets and advertising, but in general men are less likely than women to rely on the experience of their peers, preferring to live life as 'normal' (Banks, 2001; Lloyd, 2001). A study of young men's use of sexual health services found that the average young man is unlikely to access any help or support at all if he has a problem (Biddulph and Blake, 2001).

These gendered structures and practices raise interesting questions in light of the promotion of more non-face-to-face means of accessing health information and communicating with health care professionals. In the United Kingdom, NHS Direct, NHS Direct Online and information kiosks in places such as railway stations and shopping centres have all become more prevalent in recent years and some commentators on men's health have argued that the provision of more anonymous sources of health information will actually improve men's access to health care. Banks, for example, has argued that the success of the Impotence Association helpline—which receives tens of thousands of calls from men about erectile dysfunction—suggests the importance of more anonymous, confidential services for men (Banks, 2001).

When it comes to the medical encounter, gender again emerges as a significant factor, both shaping and shaped by the exchanges that take place between patient and practitioner. For example, men receive less of a doctor's time in medical encounters than women, and are provided with fewer and briefer explanations—both simple and technical (Courtenay, 2000, cited in Banks, 2001). Furthermore, although men are more likely to engage in high-risk behaviour, they receive less advice about changing risk factors than women do during check-ups. As Courtenay has found, only 29 per cent of doctors routinely provided age appropriate instruction on testicular self-examination, compared with 86 per cent who provided instruction to women on breast self-examination (2000). Furthermore, Well-Man clinics were first introduced in the United Kingdom in the early 1990s but did not take off in the same way that the earlier established Well-Woman clinics did (Kirby, 1999, cited in Banks, 2001).

Thus, there are a number of ways in which gender can be seen to shape health care practices and structures and in which, conversely, health structures and practices might be understood as contributing to the formation of gendered identities and gendered practices in health care. A parallel co-construction, we argue, is taking place in relation to the Internet.

Discussions of gender and the Internet should properly be located within the wider literature on the gender–technology relation but such studies have a long history that cannot be revisited in any depth here (for useful overviews, see Wajcman, 2004; Lie, 2003). Suffice to say that, over time, there has been a shift in focus from women to gender and from exclusion and inclusion to a more nuanced understanding of the ways in which gender and technology are mutually constitutive. The focus on gender enables analysis of how men as well as women are gendered and at the durability of the assumed link between masculinity and technology, often in stark contrast to observable variation in practice. A central concern in gender and technology studies then is the tension that often exists between, in Harding's terms, symbolic gender, gender structures and gender identities (Harding, 1986). As has been pointed out elsewhere (Henwood, 1996; Henwood and Hart, 2003), these three aspects of gender may be understood as interrelated but that does not mean that they always pull in the same direction. Thus, whilst 'gender' and 'technology' can be understood as cultural categories, used in popular discourse as a way of making things familiar and easier to deal with (Lie, 2003), we need to remain alert to the practices and processes of change that cause such symbols to be mobilized. Such mobilization may be more pronounced when traditional gender structures or identities are perceived as under threat. For example, in Henwood's study of reactions to the UK-based 'WISE' (Women into Science and Engineering) initiatives in the mid-1980s, she argued that gender binaries and essentialisms were re-asserted as a direct response to the perceived threat to traditional gender boundaries posed by equal opportunities initiatives (Henwood, 1996). In this chapter, we contend that a similar process is occurring in response to the threat posed to traditional gender binaries by the informed patient discourse.

The Internet has no fixed meaning and, in gender terms, cannot be understood in binary terms—as either pink or blue. It is constantly evolving and its meanings are constructed in use, as well as in production. As Miller and Slater have argued, 'the Internet is not a monolithic or placeless "cyberspace"; rather it is numerous new technologies, used by diverse people, in diverse real-world situations' (Miller and Slater, 2000: 1, cited in Consalvo and Paasonen, 2002: 8). Nevertheless, gender binaries continue to circulate, not only in popular discourse but in media representations of the Internet and (less explicitly perhaps) in academic discourse, too. Dorer has shown how, in the Internet's 'third phase', characterized by commercialization, the media ceased to address themselves specifically to technically sophisticated male users and hailed women as 'mothers, cooks and consumers' (Dorer, 2002: 67). In academic discourse, Sadie Plant relies heavily on this gender binary when she argues that information networks, such as those represented by the Internet and the World Wide Web in particular, are signs of feminization, offering an 'alliance of women and machines' (Plant, 1997) as do those who argue that the Internet is in danger of becoming a masculine space where women are successfully 'flamed out' as men dominate online interaction and masculine priorities continue to shape the Net (Herman, 1999; Spender, 1995; Scott, Semmens and Willoughby, 2001). Others appear to think beyond the gender binary, arguing that the Internet is better understood as a transformative space where gender categories become reconfigured (see, e.g., Turkle, 1995; Stone, 1995) but even here, there appears to be more emphasis on gender 'swapping' (implying two opposite genders) than on reconfiguration as such.

Other work has attempted to account for historicity and differences between women and to relate online and offline worlds by developing what Consalvo and Paasonen (2002) have called 'situated formulations of cyberfeminism' (see, e.g., Braidotti, 1996; Wakeford, 1997). Meeting this body of work from a different starting point are those studies of gender and the Internet in everyday life that build on earlier work on the 'domestication' of technologies in media studies work by, for example, Silverstone and Hirsch (1992) and in technology studies work by, for example, Lie and Sørensen (1996). These works have been influential in our thinking and approach, as has the work of Bakardjieva (2005) and Bakardjieva and Smith (2001) on the use of the Internet in everyday life. In particular, Bakardjieva (2005) examines not just how people use the Internet but how they 'relate' to it. Following Turkle (1984) and Aune (1996) on computers and Moyal (1992) on telephones, Bakardjieva distinguishes between those people whose main concern is with what the Internet can 'do' and those who are more concerned with how it makes them 'feel' (Bakardjieva, 2005: 19). Some people have more 'instrumental' relationships to computers and the Internet and others have more 'expressive' or 'intimate' ones. The more 'intimate' the relationship, the less technology is experienced as a tool and the more as an aspect of self—something that enhances one's status perhaps or threatens one's sense of identity. This is an important contribution to the debate about Internet use and one that we explore in our narrative analysis, later in this chapter.

One further area of Internet research that is important for our analysis here is that which focuses specifically on non-users. Wyatt et al. (2002) have offered an important insight here by arguing that not all non-use is involuntary (as is implied in the much-used term 'digitally excluded'). By so doing they have re-introduced the notion of agency into digital divide debates. Selwyn's analysis of those who stand 'apart from technology' has also emphasized the importance of human agency in non-use of technologies (Selwyn, 2003). In his article addressing non-use of ICTs in everyday life, he offers an account of non-use (and indeed all variations of use) that emphasizes the importance of the 'reading' of (or meanings attributed to) technologies. Acknowledgement of this symbolic aspect of technologies was also used in one of the (still) few studies of gender and ICTs in health care to explain midwives' resistance to electronic patient records (EPRs). Here, it is argued, computerized record-keeping was read by midwives as 'masculine', in contrast to the more 'feminine' work of midwifery and it was in these terms that midwives sought to rationalize their resistance to EPRs (Henwood and Hart, 2003). In this chapter, we continue to explore the symbolic aspects of technology by analysing how the Internet is 'read' by our participants in the context of a study about Internet use and the informed patient.

Re-introducing agency into the debate about non-use of technologies should not, in our view, be done at the expense of an account that recognizes structural factors shaping use and non-use. Indeed, just as with our understanding of gender where we seek to explore the relationship between structure and agency, technology's 'duality' has also been recognized (Orlikowski, 1992). The choice to use or not use technologies is clearly not a free one and structural constraints and enablers can be seen to exist in, for example, institutional hierarchies and domestic divisions of labour. Indeed, structural factors will make some symbolic 'readings' of technologies more likely than others and it is our contention, that, in order to better understand how gender shapes Internet use, we need always to keep in mind the relationship between structures, symbols and identities.

Counting men and women, constituting gender

The data on which this chapter draws was collected during a study that examined the health information practices of a group of mid-life and 'older' men and women who were recruited on the basis of two gender-specific health problems—erectile dysfunction (ED) for men and menopause for women.[1] We recruited those who were considering using, were using or had used the two most conventional treatments for these conditions at the time—Viagra for men and hormone replacement therapy (HRT) for women. We examined how these men and women came to a decision about whether or not to try these treatments. We sought to identify their main information sources and the media used to access these sources and their deliberations regarding the associated risks and benefits of taking or not taking Viagra and HRT. The aim was to examine the significance of the Internet to these people in the context of their wider health information landscapes, their decision making regarding health care and treatments and in their relationships with health care practitioners.[2]

It is important to remember that our study was of older women and men and there is some suggestion that age may now be the strongest 'digital divide' but this, in and of itself, does not render gender insignificant. Liff and Shepherd (2003) have analysed data from the Oxford Internet Survey (a representative sample of 2030 respondents in May and June 2003 in England, Scotland and Wales) and have come up with a number of interesting gender-related findings. For example, they found that, of the retired population, 25 per cent of men but just 18 per cent of women used the Internet; that broadband access is higher amongst men; that men have more access to the Internet at home (92 per cent men and 86 per cent women said home was their current access place) and that, even after controlling for length of experience, women were still less confident than men in using the Internet. Studies of the 'lived experience' of Internet use can offer insights into the complex workings of gender that lie behind such figures.

A simple count of men and women's use of the Internet for a range of activities can tell us very little about gender when the numbers are so small and when a study has been designed as a qualitative one. However, we present the numbers here because they suggest a number of lines of enquiry that we were able to follow up through detailed qualitative analysis of the interview data. Our number counts showed no major differences between men and women in relation to digital connection (about two-thirds in each case had access to the Internet somewhere convenient, namely home or work) but use figures were slightly lower for women than for men (20 of 24 connected women used the Internet compared with 9 of 10 connected men). More interesting perhaps are the figures for use of the Internet specifically for health-related activities. Here, women's use was higher than men's—18 of the 20 women users used the Internet for health-related activities, compared with only 3 of the 9 male users. These use patterns make sense in terms of what is already known about gender and health where women take the major responsibility for health in mixed gender/heterosexual households and where health is still generally framed as a feminine area of interest and responsibility.

An interesting set of questions arises when looking at the figures for use of the Internet in relation to ED/Viagra and menopause/HRT, specifically. Here, men who used the Internet for health-related activities also all used it for ED/Viagra-related material whereas just 10 of the 18 women who used the Internet for health-related activities had used it for HRT/menopause-related material. It is difficult to know how much to make of these differences but several explanations for women's non-use in this context are suggested by further analysis of the interview data. First, it did seem that when it came to menopause and HRT-related information and advice, women seemed to find what they needed through traditional sources—GPs, but also family, friends and work colleagues and traditional media, such as women's magazines and the health pages of daily newspapers. Second, a significant minority of women not using the Internet for this purpose were those who had originally taken their doctors' advice to start taking HRT and who had experienced no problems with it—they therefore could be construed as simply having no 'need' for further information and hence no need for Internet use in this context. Finally, as we discuss below, for others, the Internet came to symbolize an undermining of the doctor's authority, restricting use lest this undermine the trust relationship, inhibiting access to appropriate health care at some later time.

The numbers above tell us little on their own but do suggest some interesting gender-related questions that can be better understood through detailed qualitative analysis of the situated use of the Internet in the context of health-related activities. In the next section, we illustrate how gender shapes Internet use in a health context and is, at the same time, constituted in relation to both health and the Internet. We do this through a narrative analysis, which can be understood as being co-produced by our study participants and ourselves as researchers/analysts. It is important to emphasize this co-production process. By situating our questions about Internet use firmly within the context of a set of questions about health information practices, health decision making and practitioner–patient relationships, we were clearly drawing on ideas and assumptions from the informed patient discourse, where the Internet is positioned as central and potentially transformative of these practices and relationships. Thus, whilst our aim was to analyse how people positioned themselves in relation to this discourse, we can now see quite clearly how our questions were, at times, understood as promoting the IP discourse and its associated practices, resulting in sometimes forceful statements of rejection and resistance to the Internet and, at other times, attempts at compliance with the connection imperative embedded within the IP discourse.

Narratives of everyday practice: gender, health and the Internet

We begin our narratives of everyday practice with stories from Victor and Peter. Both men experienced the 'connection imperative' implied by the particular configuration of the Internet in the informed patient discourse but responded to it in very different ways which we explore in relation to their different relationships to masculinity which are, in turn, shaped by their specific health contexts.

Victor is 59 years old and married. He and his wife have three children, all grown up, with the youngest living partly at home still. Victor works as a personnel consultant and his wife works part-time as a school secretary. He has had a low sex drive for a year and the possibility of a low testosterone count was being explored at the time of the interview. He had tried testosterone patches and Viagra that worked (technically) but he did not 'feel like sex'. The sexual counsellor thinks his problem with ED may be psychological but Victor is not keen to think about that possibility.

With regard to health information and the management of health, Victor is quite traditional. He trusts doctors and nurses to provide him with the information he needs and tends not to look things up himself though he did once visit a 'Well-Man' clinic, which he read about on the notice board in his doctor's waiting room.

Victor is very defensive about his non-use of the Internet:

> I'm not interested in computers. I just want to get on with my life in the easiest way. I can't see what benefits it would give me. I'd rather hear it from the horse's mouth and talk to someone. Not get lost in cyberspace. Seems bloody obvious, and a waste of time doing this research, I reckon.

Victor's reaction to questions about possible use of the Internet in the future is very strong and we want to suggest that this may be linked to his fragile sense of masculinity (see Wyatt et al., 2005 for a first attempt at this analysis, with longer interview extract). If, as some feminist scholars have suggested, men's close relationships with technologies are forged on the basis of a sense of impotency in other areas of life—for example, personal relationships (the so-called 'hacker thesis'—see Faulkner, 2000 for discussion) or class power (McNeil, 1987)—then this defensiveness (rather than the non-use itself) would make sense. It is as if our questions are drawing attention to his lack of potency not only in the sexual arena but in the potential, compensatory arena of technology as well. We suggest that Victor's particular fragility in relation to his sexual impotency can be explained by the lack of a clear physiological explanation and his very apparent unease with a psychological one.

This interpretation can be supported in part by examining a contrasting case: that of Peter. Peter is 66 and has recently retired; he is separated from his wife. He has four sons, of whom the first and the third live with him. There is a connected computer in the house, but only his sons use it. While he feels some pressure to become a user, he says,

> I don't use it. I can't be bothered. ... I should use the computer more, I just, I don't know, I just can't be bothered. I'm lazy about that. It's not my sort of thing. I'd rather pick up a phone and talk to somebody rather than send them an email which I find takes too long.

Compared with Victor, Peter is not at all defensive about his non-use of the Internet. He talks easily about his preference for the immediacy and presence offered by the telephone, and his self-confessed incompetence in dealing with email. This more relaxed response to questions about Internet use parallels his ease in talking about his ED. Peter's erection problems are linked to his diabetes. Having a clear physiological explanation for his ED, we would argue, made him less susceptible to a sense of personal inadequacy than Victor and left his sense of masculinity rather more intact, making his minimal interest in technology and the Internet easier to voice.

We now turn to examine two women participants' relationships to health and the Internet. We start with Barbara whose account demonstrates extremely well the interconnection between gender structures, gender symbols and gender identities, and then we meet Kathy, whose story offers important insights into how gender intersects with the symbolic aspects of the Internet to shape use practices and, ultimately, access to appropriate health care.

Barbara is 50, a single mother with two sons, the youngest of whom lives with her. She works in hospital administration and owns her house. Barbara was having a bad time on HRT, which she was prescribed after a total hysterectomy a few years prior to our first interview with her. She had an implant at first but later tried both patches and pills. She had gained weight and was experiencing severe migraines since starting on HRT. By the time we talked to Barbara a second time 6 months later, she had come off the HRT, citing the headaches and weight gain as the main reasons. She was trying reflexology for the headaches and had joined Weight Watchers. She was also trying black cohosh, a herbal remedy that some believe can help alleviate menopausal symptoms. She is generally mistrustful of doctors whom she feels do not listen to women. She is quite critical of what she sees as the 'over-prescribing' of HRT.

Barbara has Internet access at home and at work but is a very reluctant user. In her first interview, Barbara describes her feelings about the Internet:

> I hate using the Internet. I hate computers. I really hate them. I guess if I was put in a corner, I had something that was really worrying me, then I would, but that hasn't happened, thank god. And so, I hate even switching a computer on. I can't stand the things.... I'm useless with a computer... I think it's just my age group. I just hate them. They don't interest me whatsoever, frustrate me like mad. It's like having a car that doesn't work to me. I get raging with them and I don't want to be there. I'd rather pick up the phone.... can't bear the damn things. I know I've got to eventually give into it.

Here, she describes her relationship to the Internet in a way that suggests a mixture of dislike and under-confidence, which she ascribes to age. However, the rage that she clearly feels may be explained, in part, by the sense of pressure she feels to give in and use the Internet, as suggested by the sense of inevitability about use, expressed in her final comment.

Barbara also explained how her son is the main user of the home machine, a situation that one might think was not a problem for her given her dislike of computers. However, by the second interview, we learn more about what we now prefer to see as Barbara's ambivalence towards, rather than hatred of, computers and the Internet—an ambivalence that appears deeply rooted in gender structures, is clearly expressed through gender symbols and lived through a gendered identity forged in relation to her son.

By the second interview, Barbara's son has left home:

> I've had a teenage son around for years who has been hogging the Internet and he won't show me how to use it because I'm too slow. He's actually just gone to University so I have actually this week been trying to get it up and running and sort it out but it's in such a mess and I have got various neighbours who keep promising to come round and help me, so I'm getting close to it. I've actually been on a basic course now. I don't like it. I hate it. … well, I know I've got to do it. I'll learn as little as possible but as you get older it really is quite hard to take on board especially if you're not interested in it. I have no interest in it at all. I just think it's absolutely boring. The thought of switching it on is like doing the ironing to me.

Barbara's account of her relationship with computers appears to be slightly different this time. She describes her son as 'hogging' the Internet, suggesting that it was something she would have liked more access to but was somehow prevented from having whilst her son was around. Indeed his construction of her (the projected identity) as 'too slow' to teach may well be a reference to her age, something Barbara continues to draw upon in her own subjective identity construction. Between them, it seems that they constructed both gendered and aged identities for themselves and each other in relation to Internet competence. However, with her son out of the way, Barbara does, in fact, get on line. She has identified the problem with the machine's slowness—the 'junk' left on by her son—and has been on a course to learn more so she can sort the problem out. In the final part of the account, Barbara reverts to her initial positioning again—her 'hate' for computers but this time we gain a deeper insight into her strong feelings. Here, she compares computer work to domestic work—cleaning up the computer is 'boring', and 'like doing the ironing'. Her feelings about computers now seem linked to a more general resentment about cleaning up after her son.

Later in the interview, Barbara talks more positively about what the Internet could offer her—specifically in relation to health and health care experiences:

> I think it would be useful to get information from other women, their experiences of using HRT or having hysterectomies or whatever.... I think that's where it's going, probably already is. It's just that I'm so slow with the Internet, it probably is already on there but I think that would probably be the way forward, networking through the Internet... That's the best place to get it [information] from—other women—not from doctors who are influenced by drug companies.

Here, Barbara can envisage overcoming her ambivalence about computers in order to take control over her health by sharing experiences with other women. She still refers to her 'slowness' but is willing to keep on trying in order to network with other women who she sees as good sources of information on HRT, hysterectomies and so on. When she makes a direct comparison between information from other women and from doctors, who she sees as influenced by drug companies, Barbara can be understood as drawing on understandings of the Internet as somehow 'transformative', enabling a challenge to medical expertise, going beyond the more neo-liberal notion of 'empowerment' found in the IP discourse.

Barbara's story illustrates quite neatly the way in which different understandings or 'readings' of technologies can influence use. When she reads the technology as a machine, full of junk left there by her son for her to clear up, it seems to signify work, domestic work even, and Barbara presents herself as a reluctant user. However, when she reads it as a communication medium, a means to network with other women with shared health problems, she becomes more enthusiastic about Internet use.

We now turn to Kathy's story which illustrates clearly, as did Barbara's, not only how gender intersects with the symbolic aspects of the Internet to shape use practices, but also how this process may be linked to adverse material consequences by limiting Kathy's access to appropriate health care.

Kathy is 57 and has been married 35 years. She and her husband have two grown-up children. She left school at 15 and has been working as an auxiliary nurse for 30 years. Her husband is a steel fixer. They own their house. Kathy was first prescribed HRT for a hormone imbalance. She took it for a few years then stopped because she was due for an operation on what was thought to be an ovarian cyst. It turned out to be a twisted fallopian tube. She went back on HRT after the operation. She was a very reluctant user of HRT and would have preferred to go through the menopause 'naturally'. She was concerned about side effects and worried about whether she was doing the right thing. However, Kathy's relationship with her GP was fraught as he thought her to be 'neurotic', a label that, although rejected by Kathy, nevertheless seems to have affected her future information seeking practices.

When we first met Kathy, she had Internet access in her home but did not know how to use it. Her husband was the main user. However, when we spoke to her again nearly a year later, she had been on a computer training course that included Internet training and had really enjoyed it. She was planning more training. By this time, Kathy was using the Internet for email and shopping and had looked for health information on behalf of her sister but not for herself. This seemed strange given her interest in health and her ongoing concerns about HRT use. Exploring this further with Kathy, we concluded that her relationship to the health care professionals and their tendency to label her as neurotic was actually working against her attempts to use the Internet to become more informed.

Kathy had a very real fear about how her questioning of the doctor would be interpreted and the effect this would have on her access to appropriate care. In particular, she feared that if she refused the doctor's advice to go on HRT, she might be refused treatment if she developed osteoporosis:

> I was still very anxious about being on HRT and all I wanted to do was to come off it but when you read the literature you're given in the packet it said you must consult your doctor ... and whenever I tried to consult my GP, they didn't want me to come off. I think he [GP] had been to a seminar and he'd been brainwashed into thinking it was the best thing since sliced bread ... I didn't want to be on HRT, I didn't want to upset him because he told me that the reason he wanted people to go on HRT was for preventative medicine to prevent against osteoporosis and heart problems and the way the NHS was going, if

people didn't look after their own health, they would become too expensive to treat.

Kathy's interest in health and in the active seeking of health information to support health choices would seem to fit the ideal model of the informed patient and, in so far as it involved taking responsibility for health matters, was clearly in line with traditional notions of femininity, too. However, her attempts to engage critically with her GP about the risks of HRT were dismissed as neurosis, effectively limiting the effectiveness of such attempts. Here, we can see how gender was mobilized (the labelling of Kathy's practice as neurotic) by the GP in the face of the perceived threat to his authority posed by Kathy's critical questions. This, in turn, can be seen to have restricted her Internet use, for fear that such use would be read by her doctor as part of that same critical practice he resisted.

Conclusion: all change, or does it?

In this chapter, we have shown how the Internet's symbolic significance cannot be underestimated. All participants can be understood as taking up positions in relation to the 'informed patient' discourse, within which the Internet figures as a key actor and is understood as empowering, if not transformative. This is not entirely surprising and we have already acknowledged the very significant role played by our research, its framing and its questions, in this regard.

There is evidence that both women and men feel the pressure to become more informed in relation to health but this posed more problems for men than women, especially in traditionally gendered households where women take responsibility for health. In relation to the Internet specifically, we have shown that the masculinity-technology equation was evident, even though it was not always expressed explicitly. Thus, we have argued that Victor and Peter's responses to questions about Internet use can be understood as indications of their awareness of this equation and that the different positions they take up in relation to it can be explained in part by their particular experiences of ED and the differential impacts these had on their masculine identities.

Barbara's story illustrates how the use of and relationships to the Internet are forged in relation to significant others, in this case her teenage son. Here, gendered and aged identities were constructed by each of them in relation to the other and it was not until Barbara's son had left the household that she was free to explore a less limited and limiting gender subject position in relation to the Internet. However, even then, so long as her son's 'junk' remained on the computer, her work in cleaning it up is resented by her and she still presents herself as a reluctant user, justifying that reluctance through reference to housework and ironing, clearly mobilizing gendered symbols and meanings (Harding,1986; Lie, 2003).

Kathy's story, as with Barbara's, shows how women can, and do, move from being non-users to users of the Internet and how, as in Kathy's case, they can become quite comfortable in its use in certain contexts. However, it also illustrates the way in which, in the context of her own health care experiences and practices, specifically, the mobilization of gender symbols by her GP inhibited her active information seeking and her use of the Internet. Kathy believed that Internet use was being read by her GP as a type of critical practice that could threaten his authority. This belief, whether justified or not, can be seen as limiting Kathy's Internet use and her active approach to information seeking more generally.

The analysis offered here demonstrates, once again, the value of studies which explore the lived experience of Internet use, rather than simply employ counts and measures of Internet usage patterns. The specific context of Internet use is always important for understanding such usage patterns and we have shown how understandings of gendered use of the Internet in the context of health information and decision making need to take into account not just structural and material factors such as divisions of household tasks (including responsibility for health of family members) and the skills and competence of the potential user, but also the more symbolic aspects of use and the way this links to identity or a sense of self. We have taken this analysis further and suggested that these different engagements with the Internet need to be understood within the wider discourses within which such use practices occur. We have argued that, in the context of this study, the discourse of the informed patient, with its imperatives about becoming informed and using the Internet to do so, is key to understanding the gendered patterns of use we observed.

While the accounts given in interviews suggested a range of different levels of engagement with the Internet by both men and women, they also re-asserted the gender binary in ways that suggest that the informed patient discourse constituted a threat to traditional gender boundaries in the areas of health and technology. Some circumstances, practices and discourses change, such as the introduction of the Internet into many people's daily lives and an emphasis by health policy makers and practitioners on the need for patients and their carers to be informed. At the same time, some practices and discourses, in this case those around gender, remain obdurate and resistant to change, and may actually become stronger if seemingly under threat. It therefore remains to be seen how the informed patient discourse will be worked through in both a gendered and a digitized context.

Notes

1. The study, entitled 'Presenting and interpreting health risks and benefits: The role of the internet', was undertaken between 2001 and 2003 and was funded under the joint UK ESRC/MRC research programme on Innovative Health Technologies (www.york.ac.uk/research/iht).
2. Thirty-two women were recruited via a family doctor or gynaecological clinic and 15 men were recruited via a urology clinic, a psychosexual counselling service for men suffering from erectile dysfunction or via a diabetes clinic. All were interviewed once and about half were interviewed a second or even third time in the period between November 2001 and January 2003. All first round interviews were audio-recorded and fully transcribed. Some follow-up interviews were conducted via telephone and email. The follow-up interviews provided a longitudinal dimension to the study, providing us with an opportunity to focus on changes over time, in Internet use as well as in health conditions and treatments. Of the 32 women interviewed, the average age was 55, with the youngest being 39 and the oldest 73. Eighteen were in a relationship. The men were older, ranging from 54 to 81, with an average age of 66. Eight were in a relationship. As far as we are aware, our sample included only heterosexual women but included a mix of heterosexual and gay men. The sample as a whole included a range of socio-economic groups, with varied educational experience and qualifications. Most were white and British.

2

Gendered Identities and Caring: Health Intermediaries and Technology in Rural and Remote Queensland

Lyn Simpson, Michelle Hall, and Susan Leggett

Rural and remote communities experience significant hardship in terms of the ready availability of high level emergency health care, access to resources, supportive technology, and high staff turnover, resulting in gaps in both initial service provision and ongoing support. In this context, local people with health knowledge or caring skills can play an important bridging role. Non-clinical health workers, allied and community health, social workers, and other community members may work to fill the provision gaps, providing information and support to community members, in many cases over and above their paid responsibilities. These people can be defined as health intermediaries; they act as mediators between the health system (or lack thereof) and the individual needs of community members, providing information and support in ways that best suit the individual and his or her specific context.

In this chapter, we draw on our research into the information requirements and supportive behaviours of health intermediaries in rural and remote communities in Queensland, Australia. We examine current assumptions about the usefulness of new information and communication technologies (ICTs), primarily the Internet, in facilitating the provision of health information to community members, in the light of the importance of social support for encouraging healthy behaviours and the traditional association of women with this supportive role. Using examples from our research, we then examine this social support role in a rural and remote context, exploring how a gender stereotype can translate into assumptions regarding the requirements of the health intermediary role, and for supporting technologies and the contradictions this presents in the light of e-health policies. Our findings, especially in very isolated areas, point to health as a complex and personal issue that is firmly situated in the social, rather than the technological realm. The practical and effective on the ground health care practices we saw occurring in rural and remote Queensland raise questions about assumptions regarding the roles for technology and women in the provision of health information and care, and what we see to be the separation of government approaches to health information and health care provision.

Health information and the informed patient

The delivery of health information through ICTs is seen by governments as an important way of extending and strengthening the provision of health services, especially in rural and remote areas where health infrastructure may be limited (see, e.g., National Health Information Management Advisory Council, 2001; National Rural Health Policy Sub-committee and National Rural Health Alliance, 2002). For governments and health providers, ICTs offer potential for the cost-effective delivery of a wide range of up-to-date information in a timely, convenient, and relatively private manner (Department of Finance and Administration, 2005; Gaby and Henman, 2004; National Health Information Management Advisory Council, 2001). Through easily accessible health information, consumers are seen to be more able to seek out relevant information to assist them in making decisions about their own health and well-being (National Health Information Management Advisory Council, 2001), an approach that is compatible with government moves towards 'focusing the health and aged care system more on healthy lifestyles, prevention and early intervention' (Department of Health and Ageing, 2006: part 3). This relies on the construct of an 'informed patient': an empowered individual who actively seeks out and makes effective use of health information for their individual benefit (Hardey, 1999; Nettleton and Burrows, 2003). Viewed as a preventative mechanism, the individual as informed patient takes greater personal responsibility for their health, and reduces their reliance on the formal medical system. This potentially has major cost benefits for governments, as well as illness-prevention benefits for the population.

The notion and effectiveness of the 'informed patient' construct, however, rely on considerable assumptions regarding an individual's ability and willingness to engage proactively with health information and ICTs. Issues with access due to the 'digital divide' are routinely raised (Loader and Keeble, 2004; Marshall, 2004; Wyatt, Henwood, Hart, and Smith, 2005) and are significant in our research site. But other concerns, such as levels of general, computer, and health literacy (Kickbusch, Wait, and Maag, 2006), usability of health websites (Moon and Fisher, 2006), information bias, information overload and self-diagnosis (Fox, 2006; Mackian, Bedri, and Lovel, 2004; Nettleton, Burrows, and O'Malley, 2005), and a general unwillingness to engage in informed patient behaviour (Henwood, Wyatt, Hart, and Smith, 2003) have all been raised through research. Despite these concerns, much e-health policy still relies on the concept of an active, literate consumer and focuses on meeting their information requirements, paying little attention to the ways such information, once accessed, is actually used by consumers. The provision of information through ICTs is privileged over the skills required to properly reflect on and understand this information, skills which are often ignored or taken for granted. This conflating of the idea of information provision with understanding (Brown and Duguid, 2002) ignores other contributing elements, such as support from social networks, that have also been found to play a role in affirming, supporting, and rewarding good health practices and may have a mediating influence that helps encourage reflexive behaviours (Lupton, 1997; Preece and Ghozati, 2001). We need to recognize that there is more to encouraging informed patient behaviour than merely providing end users with access to health information (Wathen et al., 2008). It is only by exploring, more closely, the processes that encourage reflexivity, such as social support, that we can more effectively consider the ways in which ICTs may contribute to this process (see, also, Bella, this book).

Health information and social support

The importance of social support and social networks for maintaining individual and community health and well-being is well established in mental health and public health research (Brennan and Fink, 1997; Gottlieb, 1981; Israel, 1985; Israel and Rounds, 1987). Support from social networks is considered vital in assisting those with low health literacy to deal with the health care system by allowing for individual and cultural contexts to be made relevant (Kickbusch et al., 2006; Lee, Arozullah, and Cho, 2004). Informal helping systems activated through social networks are also viewed as a means of complementing the formal health care system through their ability to communicate with the wider community at a 'lay' level, and to serve as a bridge between health agencies and the community (Eng and Parker, 2002; Mitchell and Hurley, 1981). This is especially relevant in rural and remote communities where health agencies may have a limited on-the-ground presence, where resources are limited and community health workers and NGO services are often the only permanent non-clinical health staff present.

One of the important reasons that social support is effective in encouraging positive health outcomes is its role in encouraging 'everyday understandings' of health information (Israel, 1985; Kivits, 2004; Lee et al., 2004). The medical terminology used in much health information (from doctors, in books and online) can be difficult for many people to understand (Simpson, Stockwell, Leggett, Wood, and Penn, 2006) and it may not be congruent with their own experiences and culturally specific understandings of health overall. For these people, social support from a trusted member of their social network can help them to translate medical information into a language that makes sense to them and their current health context. This is of significant benefit for people with lower levels of health and general literacy and limited access to medical services, who may not possess the necessary skills to engage reflexively with health information.

Social support networks effectively extend access to resources such as the Internet, through individuals who are actively involved in seeking health information for others (Madden and Fox, 2006). Health intermediaries, for example, act as mediators between 'official' information sources and end consumers, combining information and their supportive abilities to encourage understanding. In this context, the Internet may be most useful as an information resource for health intermediaries, who then translate the information retrieved into a format that is accessible to those in their social network. This is particularly relevant in rural and remote areas such as our research site, where access to Internet resources is limited due to poor infrastructure, high costs, and slow access speeds and where literacy issues associated with the digital divide are prevalent (Simpson, Hall, and Leggett, 2006).

Social support networks may be seen, then, as a means of encouraging understanding rather than simply providing information, since these supporting activities allow not just for an exchange of information, but also for emotional support (Harris and Wathen, 2007). Those providing social support act as mediators, supplementing health information with other supporting processes and the ways of their own community and culture, encouraging everyday understandings in health consumers, thus building their capabilities for more informed and responsible health behaviours.

Social support and gender: who cares?

Social support can be broadly classified into four types of supportive behaviour: emotional support (affect, esteem, concern), instrumental support (aid in labour, time, money), informational support (suggestion, information, advice), and appraisal support (affirmation, feedback) (Cwikel and Israel, 1987). We draw on these distinctions to more closely investigate the elements of the health intermediary role. In so doing, we move away from the notion that informational support is associated solely with men (and with technology), and emotional support with women (Suitor and Pillemer, 2002; Wajcman, 1991; Wellman and Wortley, 1990), a reading that might follow from Harding's notion of 'gender symbolism' (Harding, 1986).

To a large extent, health caring roles (as distinct from clinical roles) are seen to mirror domestic caring roles; both involve providing physical support, comfort, nurturing, and non-specialist information to unwell or needy others: children, and unwell adults or elderly people in both formal and informal circumstances. Health carers are typified as having particular qualities, such as kindness, consideration, sympathy and tact, patience, reliability, and unselfishness (Skeggs, 1997), and these qualities in health caring have come to be synonymous with women (Poole and Isaacs, 1997). When technology is added to the mix, these assumptions may be complicated, with masculine (and bureaucratic) concepts of the provision of information through technology superseding other (traditionally female) supportive requirements of health care (Wajcman, 1991). However, appraisal, emotional and instrumental support are important to consumers and those who support them (Henwood et al., 2003; Kivits, 2004) and masculinized notions of technology in health caring may overlook this dimension.

A number of authors have sought to move beyond this simplistic gender binary, adopting the notion of femininities and masculinities (Coppock, Haydon and Richter, 1995; Fitzsimons, 2002; Kenway and Fitzclarence, 1997; Poole and Isaacs, 1997; Seidler, 1998; Skeggs, 1997). This formulation allows for a distinction between sex and gendered roles, since the emotional caring dimension can also be carried out by men, and managerial, informational or technical/scientific elements by women. While the parameters of these identities are in flux (Paechter, 2006), nurturing and caring behaviours are most often discussed in terms of femininities (just as emotional and appraisal support are linked with women), whilst technical, leadership or expert roles are seen as displaying masculinities (similarly, informational support is seen as more likely to be provided by men). The relevance of the masculinities and femininities framework to health care work has been recognized in research by Azuma (1990), who found that successful career nurses, defined as those with long service and who had been promoted to a senior position, scored higher in masculinity than others. Hirokawa, Yagi, and Miyata (2002) also support this position, suggesting that within the nursing and social work professions, feminine traits such as empathy for others and masculine traits such as independence and assertiveness may both be necessary for better care-giving. Denying the presence of masculinities within health care work ignores the importance of problem-solving skills and self-sufficiency, downplaying significant elements of the support roles that many workers undertake. We argue that it may also have consequences for the understanding of these roles by senior bureaucrats and the resulting allocation of resources, such as ICTs.

Adopting this framework of femininities and masculinities, and recognizing their relation to the four aspects of social support, allows us to move away from the men/information/technology and women/emotional care stereotypes and recognize the fundamental social support practices that make the health intermediary role so effective in rural and remote areas. It also provides a more useful framework in which to explore how the focus on ICTs in health can both enhance and limit these supporting practices and their ultimate goal of encouraging everyday understandings of health.

Health information and health care through social support

Applying the framework of social support, and recognizing its mediating role in the provision of health information and care, we can see that informational support is only one element in providing health care. Despite this, there is a focus in Australian government e-health policies on driving greater individual responsibility for health via health information provision to the end user, reinforcing assumptions that the availability of information automatically leads to understanding. As a result, other aspects of supportive health care work are undefined, or perhaps taken for granted, reinforcing the split between (masculine) health information and (feminine) health care.

Yet health information that is provided in a manner which is bereft of 'meaning, judgement, sense making, context and interpretation' (Brown and Duguid, 2002: xiv) and instead 'dumped' on consumers whether through the Internet, mass media vehicles such as television, or advertising, can potentially create considerable frustration and distress, and increase marginalization and the disempowerment of these people (Simpson et al., 2006). Information alone does not necessarily offer the potential to actively generate value and contribute to society, and information seekers may require the assistance of others to provide the contextual information that assists in the learning process. As Johnson (2004) has noted, people most often choose other people as their preferred source of information, for reasons such as ease of access and trustworthiness. This allows for the provision of appraisal, emotional, informational, and instrumental support, helping to ensure that health information is not only found, but that it makes sense within an individual's current context. We suggest that it is this active involvement of members of a social network in the health situations of others that has the greatest potential to spread reflexive behaviours and encourage community well-being especially in disadvantaged and under-resourced populations. If this is the case, then online health information resources that do not take into account the full range of support mechanisms nor the range of gender identities exhibited in supportive practice may be complicating the work of health intermediaries and limiting the convenience and cost-efficiency of online health resources. This is especially the case in rural and remote areas, where existing digital divide issues would seem to undermine policies that rely on ICTs to address isolation and under-resourcing. The remainder of this chapter explores these ideas through the work of health intermediaries in rural and remote Queensland, focusing specifically on the ways in which gender, technology, and social support intersect to enable or inhibit effective health information and care provision.

Exploring the practices of health intermediaries in rural and remote Queensland

Our research examined patterns of communication about health and well-being in rural and remote communities in Queensland, Australia, to explore the ways that knowledge about health is created, shared, and communicated by health professionals and community members in more isolated areas. Four sites were visited specifically for this project: an accessible town, a moderately accessible town, a remote town and a very remote Aboriginal community (see Figure 2.1).

Community health concerns varied according to the ethnic make-up of the town and proximity to major centres. In accessible towns, chronic conditions, often related to being overweight or obese, were

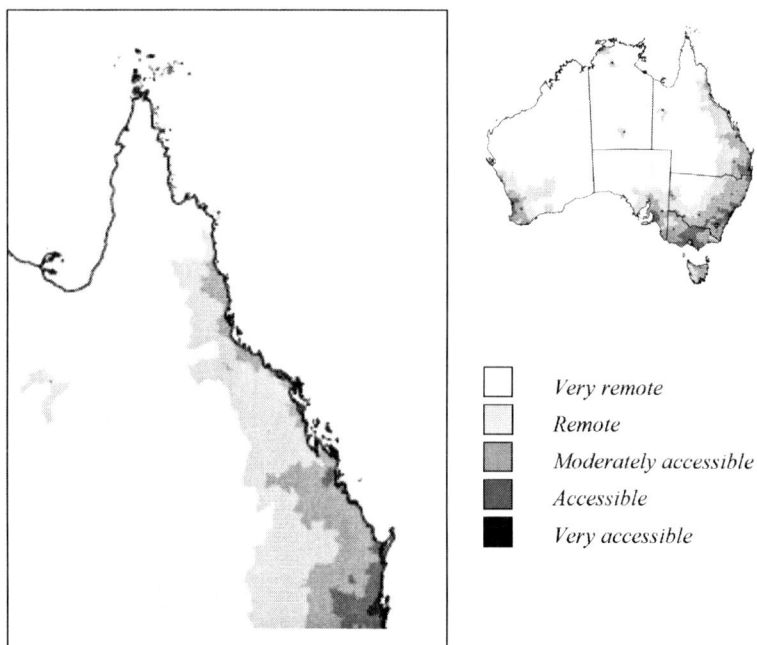

Figure 2.1 Remoteness scale, Australia.

Legend:
- Very remote
- Remote
- Moderately accessible
- Accessible
- Very accessible

Source: Adapted with permission from the Public Health Information Development Unit, University of Adelaide (2001).[1]

repeatedly mentioned as health concerns. High blood pressure, heart disease, and diabetes were common. In remote communities, where Aboriginal people make up a larger proportion of the population, these issues were present at higher levels, as was rheumatic fever. Social health problems—sexual and domestic violence, drug and alcohol abuse—were also prevalent. All communities, however, shared a concern that health resources were somewhat sparse and difficult to access. In addition, services were often seen as not appropriate—or responsive—to their rural, remote or Indigenous community setting.

We began by seeking an understanding of the ways in which people seek and share health information in rural and remote communities, surveying community members in two rural communities via a face-to-face, random street survey.[2] No matter what originating information source they identified, people indicated that they do not just require information about their health, as assumed in the informed patient construct, but also seek support in understanding that information and applying it to their everyday experiences of health and well-being. That is, the provision of information alone was not sufficient; other forms of support—emotional and appraisal—are necessary.

It became clear during the survey that those people who most often provided help and support regarding health matters were already operating in the health system, or community welfare organizations, in paraprofessional roles. This may be a function of the rural and remote context, where small populations and workforces mean that those interested in helping others with health matters find employment within the health and community services system. These workers became the focus of a second round of data collection (conducted from October 2005 to March 2006) using interviews and focus groups. The 40 health intermediaries we spoke to included Aboriginal health workers, mental health and social workers, people employed in aged care and disability advocacy, community health workers and school health nurses.

Who are these health intermediaries?

Our informants had come to health caring from a variety of roles—for example, one had previously worked in hairdressing, one as a real-estate agent, and another had worked as a stockman, and for an architect. They brought a wide range of life experiences to their roles, often noting that they felt their backgrounds enabled them to more effectively operate within the rural and remote environment:

> Disability worker: In my experience, very few community workers who work for non-government services up here, government funded services, have formal qualifications unless the funding requires that they have a psychologist or a social worker, but otherwise most of us are people who stumbled into the work and have enjoyed it and had to struggle. For me it has been very difficult...
> Advocacy worker: Basically what we are saying is, the thing about it is, we enjoy doing it and that's what makes it different.
> Advocacy administrator: It makes the whole difference...
> (remote town)

In addition, many had self-selected, responding to a perceived need in their community:

> I have been a health worker since '93. I actually went to [very remote town] and started the clinic out there from an air-conditioned

donga,[3] and... I had no training. I had no training or experience, except my interest in health from very young, and just my experience as a mother. I got it up and running, and then I went from there and did my health workers' training, in [major regional city].

(Aboriginal health worker, remote town)

In each case, health intermediaries identified their ability to network and communicate with others and their desire to help as being major personal drivers of their supportive behaviour, and as essential in the health intermediary role. Communication through social networks was vital here, both in passing on and receiving information about others' health. For Aboriginal health workers, the well-being of their people was a particular driver: 'I really want to help our people. They're the ones that are at the disadvantage' (Aboriginal health worker, remote town). This attitude resulted in high levels of commitment to and flexibility in their helping behaviour. This commitment was most evident in availability, with many health intermediaries noting that they were approached on both a formal and informal basis at any time of the day or night to offer a wide range of support and advice:

I mean it's every day, it doesn't just stop during the day, it's after hours, it's just continuous stuff. And it's not just for myself as team leader but all the other health workers here as well, when they socialize, even at home it's just continuous information given to them... you're always on the ball.

(Aboriginal health worker, very remote town)

Other instrumental behaviours included seeking or checking on community members with known health problems if they had concerns regarding their well-being or needed to pass on information, and booking or driving community members to appointments. On some occasions this could be seen to have occurred within the context of an intermediary's working responsibilities, but often such helping behaviours occurred outside of work hours and work premises and constituted social support.

How do they provide health information and care?

A major part of a health intermediary's formal and informal role included finding and translating health information into a format that is suitable for the individual needs of the person they are supporting. This translation process is seen as vital, especially for information that has come from 'official' sources, such as government health organizations, the medical profession, or the Internet:

> Researcher: And do you find that you need to—say if you get some-thing off the Internet—do you need to change the language?
>
> Aboriginal health worker: That's right. We do that quite a bit here with a lot of our stuff, a lot of our health promotion stuff. We've developed them to suit the way our people think... yeah ....
>
> (moderately accessible town)

As this example shows, a key element in this worker's role is making the health system and information appropriate for the users. This goes beyond the straight provision of information, indicating emotional and appraisal support and a concern that those they are helping not only receive beneficial health information, but also that they understand it. This was a defining feature of our respondents' practices, their deter-mination to access and disseminate health information to those who need it—in whatever way they can best do so. Successful health inter-mediaries (those who felt they were trusted by their clients and able to communicate effectively with them) noted the importance of trust, flexibility, and opportunism in their quest to do this.

Trust was considered a vital part of their helping role and interme-diaries devoted considerable effort to gaining the trust of those they were trying to help. As one advocacy worker noted, 'The way I go into the community—if [it's] someone I don't know—it's probably five or six visits before I even try... And before they'll even open up to you!' (Remote town). Without the trust of the community members health intermediaries felt that their efforts to help would be compromised or less effective, which is an issue of major concern for those working in Aboriginal populations where health outcomes are poor and relations with the white health system can be fraught.

Flexibility and opportunism were also ways that health intermediaries enhanced the effectiveness of their role and allowed them to cross the boundaries between formal and informal care provision. Flexibility in work practices enabled these intermediaries to adapt to the needs of the person they were supporting, and opportunism enabled them to seize on any potential advantage to gather or disseminate information:

> I'll use any opportunity—the football training in the afternoon—if something's happened at school and I've got notes or I need to see a parent, I know they're going to be down there and I do the thing down there. Get the information out down there, if I run into

somebody up town and I've got the form in my bag, I'll pull it out and get Mum to sign it there.

(School Aboriginal liaison worker, accessible town)

Health intermediaries serving Indigenous communities spoke of meeting clients under trees and on riverbeds, of taking AIDS education campaigns to camp fires at night, approaching people at football practice, or at the pub. They acted as transportable information resources. In many cases these dissemination practices were seen to be at odds with the standard practices of their employer, but were chosen because they offered the greatest potential benefit to the community members requiring care. What is important here, for both community and intermediary, is that the support is provided in a way that best suits the person *receiving* the help, rather than the organization or person *providing* it. That is, these health intermediaries recognize that in some cases information provision is not sufficient, and that people need support to help them to engage with it in a meaningful way. This is especially the case with Aboriginal health workers and others working in Aboriginal communities who face additional barriers of a significant cultural disconnect between Aboriginal understandings of health and well-being and the scientific definitions of the white medical system.

Is gender important?

As we continued to talk to health workers and carers, stereotypical portrayals of health caring seemed increasingly inappropriate, and we sought more fitting ways to explain the mix of attributes these health intermediaries brought to their roles. This led us to both the social support and masculinities and femininities frameworks, as a means of understanding the depth of support workers provided and the flexible ways they approached providing it. Even accepted variations from gender stereotypes (such as associations of caring with homosexuality and assertiveness and strength with rural women) did not explain the range of skills and practices we found amongst health intermediaries as they negotiated both the requirements of their communities and the lack of resources with which they had to operate. In our research we met many intermediaries in health care roles who displayed masculine traits in their supporting practice, specifically assertiveness, self-reliance, instrumental skills, public knowledge and rationality:

Since I've been you know like team leader I've been thrown in the deep end, but I've taken it on because we had people coming in and out all the time as team leader, and um it was people from [elsewhere] coming in to [the community] and they had all these different changes and stuff like that....and one of the things that I've found that we have all these specialist clinics come in, we have all these people come with all the specialities and professionalism and stuff, and yet they cannot deliver to our people...So the reason why I have done the health workers' forum is that we need to hear what the community people are saying and go with it because if we don't develop programs that's coming from the people, then nothing's going to work, they won't support it and stuff like that. So we need to acknowledge what they are saying, what professional and specialized service we have on board, and bring it back, integrate it together in such a way so that we can deliver it in a way that people can utilize the service.

> (Aboriginal health worker, very remote town)

At the same time, health intermediaries also displayed the more expected femininities such as emotional involvement, sensitivity to others, tact, and high levels of communication skills, for example:

I used to look after the kids when I was on [Very remote town] on the weekend sometimes...and those kids, they all used to sleep in my lounge on a big fold down bed I used to have, four boys and I used to have to sit there every night they were with me, Friday, Saturday, Sunday night and tell them stories about my history cause that's what they wanted to know, and they would tell me some stories about their stuff. It's such a part of their life...I think it's massively important, great story.

> (Mental health worker, remote town)

We noted that this range of characteristics was present, regardless of the worker's gender: the first example above is from a female worker, and the second, a male. Most intermediaries illustrated the ability to adapt to ever-changing circumstances by relying on masculine strength and/or feminine compassion as required. Their flexibility in moving between gendered characteristics according to their supportive role was perhaps their greatest asset in a highly under-resourced area.

How does technology fit in?

When we start to look at use and provision of technology in our research context, we begin to see how the stereotypes of a gender binary and a focus on ICT-driven information provision may impact on the availability and effectiveness of resources for our health intermediaries. Two distinct types of health information consumer were evident: what may be termed 'end users', and intermediaries. End users need information that suits their level of health literacy, that is easily found and accords with or can supplement their own everyday understandings of health and well-being. Intermediaries, on the other hand, can take advantage of resources of greater complexity that may use more specialized or expert language that they can then adjust according to the requirements of the ultimate end user. Providing health intermediaries with appropriate resources, and the capacity and tools to create and share such resources, would seem likely to most benefit their client group, since their multifaceted supporting role will enable them to provide or translate material in a way that facilitates everyday understanding. In other words, if health workers operate as information and support resources for others in their social networks, then it is important to ensure that they are properly supported with whatever resources they require to effectively do this. The significant difference between seeing ICTs as resources for intermediaries and for the end user is acknowledging the importance of the translation and adaptation that occurs between the initial ICT information provision and its final use.

However, in line with the gender binary, information intermediaries in health care roles may be stereotypically expected to focus solely on the feminine supportive elements of care, rather than the masculine informational elements (which are taken care of by doctors or by technological resources). It is possible, then, that their need for flexible resources or resource-accessing technologies may be invisible to policy makers. This may be the reason that, whilst e-health policies are promoting the importance of new technologies to deliver health care efficiencies through the provision of information to the end user, in rural and remote areas of Queensland, health intermediaries are denied access to these technologies and the information resources and networking capabilities they offer (Forster, 2005). Bureaucratic decisions to restrict Internet access to those in higher status roles, for example, mean that most publicly employed health intermediaries are unable to use this technology at their workplace. Even access to computers is severely restricted, significantly limiting their access to tools such as email and PowerPoint which might enable them to share and create suitable information resources:

Liaison worker: …I don't have that [Internet] access—[at the hospital], you've got access to the computers in the kitchen. But sometimes you don't get a chance to get in the kitchen, you know, dining room I meant...

Researcher: So do they have computers with Internet access?

Health worker: Yeah. They are always busy. When you go, the doctors are always on them.

Liaison worker: Junior doctors and doctors are always on them.

(Remote town)

Whilst this access division overtly reflects economics or professional gatekeeping, it may also relate to perceptions of technology as masculine (Henwood, 1993; Henwood et al., 2003; Wajcman, 1991) and the fact that those in care-giving roles are more likely to be women (Poole and Isaacs, 1997). Like the workers themselves, however, effective information resources and the technologies they are delivered through need to be flexible, adaptable, and always available. Because of their varied requirements for communication tools, providing health intermediaries with a limited range of technologies, or assuming that one technology (e.g., the Internet) can address all their needs, overlooks the fundamentals of the social support role, including significant flexibility, which workers need to address the community's widely varying needs.

Hence, in areas of significant isolation and under-resourcing, in terms of health and technology infrastructure and staff numbers, technologies were used according to availability and their effectiveness in terms of supporting outcomes. In many cases, it was the 'older' technologies that were most effective—their availability was greater due to familiarity, their use less restricted and their application was inherently more flexible. Technologies, such as cars, cameras, cardboard, paper and glue, were regarded by many intermediaries as being more useful than email, PowerPoint, Videoconferencing, DVDs, and the Internet, though this effectiveness was greatly impacted by context. When asked if there was anything that would enable her to do her job more effectively, one worker, in addition to wishing for more colleagues to share the load, reported that her most pressing need was for a vehicle: 'At the moment, the vehicle part is ... like getting a car to go out' (Health worker, accessible town). Whilst this particular worker had access to Internet resources, it was a technology that enabled face-to-face contact that was seen to be of greatest use in helping maintain the health and well-being of the community she served.

Even taken-for-granted technologies, such as the telephone (seen as an effective tool for networking with other health intermediaries and services), were of limited usefulness in more remote areas due to poor mobile communications infrastructure and the 'on-the-road' nature of many intermediaries' paid roles. Furthermore, as a tool for providing support to a community, the telephone was considered inadequate, not only because of limited infrastructure and take up, especially in Aboriginal communities, but also because of its limitations as a mechanism for effective support provision. As an advocacy worker noted, 'What's the good of talking to someone and counselling someone over the phone and they're watching TV?' (Remote town).

The Internet presented similar restrictions, with most intermediaries having only email access and restricted Intranet resources, and only limited opportunities to use both. The use of email to share alerts, information or resources with other workers was regarded as of little value, since most workers were unable to then access the Internet-based resources discussed. Ultimately many online resources were not considered suitable for the intended audience and those that were, were not available in electronic format. Further, networking through email is of limited value for workers who may only use a computer once a week, for short periods of time, and do not have access to printers, scanners or appropriate software for resource development and sharing.

Instead, resources developed with older technologies, such as flip charts and posters, were used in public health information campaigns. One such example is a range of graphic posters about scabies that was developed by health workers in a remote Indigenous community and disseminated both physically and online[4] using meaningful photographs from the community to reinforce relevant preventative care messages. Computer programs such as PowerPoint would more easily allow the creation, local adaptation and sharing of such visual resources, an efficiency recognized by most health intermediaries. However, the poor availability of computer technology meant that for many intermediaries this time-saving option was not available, and they were in effect recreating resources which may already have been created by other health workers in their extended network.

Given these circumstances, we can see the inconsistencies in an Internet-based information provision approach to encouraging informed health behaviour, especially in rural and remote areas. Such an approach ignores the importance of other supportive practices such as emotional, instrumental, and appraisal support and appears to operate on a gender binary in terms of how it makes those resources available. As one worker noted, there is an inherent contradiction in the fact that for many workers, 'you have the trust to look after the community's mental health, but you don't have the trust to use the Internet' (Mental health worker, very remote town). Is this lack of trust a result of the gender stereotypes of health caring roles? Is it a representation of a bureaucratic belief that those (women) who care do not require information to perform their role? Or is it simply a lack of recognition of the full range of support that non-clinical health workers provide community members in rural and remote areas on a formal and informal basis? This lack of trust from bureaucracies is in strong contrast to the intermediaries' great emphasis placed on building trust, and the corresponding value that communities place on the support provided by these intermediaries. It also points to a divide between the information focus of the informed patient discourse and the more practical approach to encouraging everyday understandings that we saw at work in communities. Given the current emphasis of governments on consumer-focused online health information, it is somewhat ironic that full access to those information resources is not provided to the health workers who play a very significant role in ensuring such information is disseminated and understood in their communities.

Conclusion

Information provision via the Internet is increasing as governments seek to cut costs; the result is resourcing policies that are tailored for highly information-focused health care provision. Much of the policy discourse surrounding technology implementation comes from bureaucracies and reflects masculine understandings of informational health support, ignoring the flexible adaptation (a feminine trait) that occurs on the ground, as users of both genders adapt both their working practices and the functions of technology to best suit their individual needs. Our research in rural and remote Queensland has demonstrated that it is this process of use and flexible adaptation through use that gives technology its potential value, especially within a rural and remote social support context. At present, where workers' ICT needs are poorly provided for, older technologies assume greater importance. Being able to develop targeted flip charts and posters that can be taken to every location—including the river bank or the camp fire—and having transport to get to those locations means that essential emotional and appraisal support can be provided face-to-face. But workers do need ICTs, since the vital informational element of their work can be simplified by ready access to the world's health information resources, and to the in-office tools to adapt these resources to a lower tech delivery.

Perhaps the key question that arises is why are these workers under-resourced? Is it because their work is seen as low-tech, hands-on, and 'women's' emotional work—which is perceived by bureaucracies as requiring limited resources—reinforcing the split between health information and health caring roles? All workers we talked to displayed a mix of femininities (associated with the caring elements of the work) and masculinities (associated with the advocacy, adventurousness, and instrumental elements) and emphasized through their intermediary practices that both were required to effectively support others. Ideally, a gendered stereotype of the health intermediary role would not result in different treatment and perhaps this is a phenomenon accentuated by remoteness; however, perpetuating that stereotype devalues the full range of supportive mechanisms and traits (masculine and feminine) these workers bring to their role and demonstrates the limited value given by policymakers to the social dimensions of health. Denying this flexibility could be seen to be disempowering for these people, ignoring the vital role they play in building and maintaining community health and well-being. Complicating their ability to provide support by denying technology's ability to assist can be seen as compounding the problem, resulting in frustration, anger, burnout, and the loss of valuable community support networks in already under-resourced areas.

Acknowledgements

We are extremely grateful to all the participants in our fieldwork, and most particularly to the Aboriginal people who were generous with their time and knowledge; we are acutely aware that they are the subject of a vast amount of research, and can feel that this is a one-sided process. We hope that this is not the case for the work we undertook with them. We are also very grateful for the assistance and generosity of the Royal Flying Doctor Service in liaising with communities and getting us to some locations, and to the Social Sciences and Humanities Research Council Canada.

Notes

1. 'Highly accessible—locations with relatively unrestricted accessibility to a wide range of goods and services and opportunities for social interaction. Accessible—locations with some restrictions to accessibility of some goods, services, and opportunities for social interaction. Moderately accessible—locations with significantly restricted accessibility of goods, services, and opportunities for social interaction. Remote—locations with very restricted

accessibility of goods, services, and opportunities for social interaction. Very remote—locationally disadvantaged—very little accessibility of goods, services, and opportunities for social interaction' (Hugo, 2001: 3).

2. The survey instrument was adapted from similar work conducted in Ontario, Canada, as part of the ACTION for Health project. A further telephone-based survey was conducted in the same and additional communities in Queensland, Australia.

3. 'Donga' is used here to refer to a small, and often temporary, building.

4. Available at: http://www.healthinfonet.ecu.edu.au/graphics/graphics_health/ wadeye_slide/healthy_skin/hs_slideshow.php.

3
Geeks Who Care: Gender, Caring and Community Access Computers

Leslie Bella

Caring work, both professional and informal, supports the social networks that are key determinants of health. Caring theory reveals this work as gendered. Direct caring work, generally understood as women's work, has been devalued and even rendered invisible. Men have been more involved in more visible executive caring. While direct caring is likely to be women's work, technology has generally been associated with men and masculinity. This chapter shows how these two contrasting gendered processes of care and technology interweave at a particular community-based setting. The MacMorran Community Centre is located in a low-income neighbourhood of St John's, Newfoundland,[1] and offers programmes to enhance the health and well-being of residents, including a programme of public access computers. Interviews with 25 key informants allowed us to explore the intersection of gendered processes involving technology on the one hand and caring on the other. We focus on 'executive care' (a more stereotypically masculine activity), 'direct care' (more stereotypically feminine) and receiving care, and on perspectives on the technology that may be intrinsic (more stereotypically masculine perspective) or instrumental (more stereotypically female). Our analysis reveals more involvement of men in caring (both executive caring and more visible, direct care), a development that appeared to enhance the community's abilities to support the health of its members. Concomitantly, the development of more technological expertise among women enhanced their potential to be effective volunteers supporting the health and well-being of their neighbours, and also led to education and/or employment opportunities. These processes associated with caring and with technology interwove to produce a generation of skilled individuals whom we have named 'Geeks who Care'.

As this chapter demonstrates, and as is echoed in Simpson et al.'s chapter in this book (see Chapter 2), realities are more complex than these stereotypes suggest.

Key determinants of health and MacMorran Community Centre

Canada's Public Health Agency (2006) uses a population health perspective identifying key determinants of health to design programmes to strengthen those factors associated with improved health status. The five most important determinants of health are income and social status; involvement in social support networks; education and literacy; employment and good working conditions and social environments containing social support networks (CPHA, 2006). The MacMorran Community Centre, established in 1983, works to reduce the negative impact of these determinants on the lives of residents, and thereby enhance their health and well-being.

In St John's, Newfoundland, MacMorran is set among 151 provincially owned housing units. Residents live on very low incomes from a combination of government support and low-wage employment. Most have limited education, and many have issues with basic literacy. They share many characteristics with the rural residents served in Simpson et al.'s Australian study (see Chapter 2, this book). The Centre's programmes include a food bank, homework support, high school completion opportunities and programmes to help people find work. The Centre enhances local networks through recreation opportunities, invites local participation as both volunteers and as part time workers and contains a federally funded CAP (Community Access Programme) site providing free access to computers and the Internet. The CAP site is central to Centre programming in relation to all five key determinants of health.

Our research was funded by ACTION for Health with MacMorran as a community partner.[2] The Centre approved the project and provided ongoing consultation. We documented the history of the Centre's public access computers (Bella, McDonald and Walsh, 2004), and worked to understand 'capacity development' as used at MacMorran (Bella and Bishop, 2004). We then interviewed 25 key informants about the role and impact of the Centre's public access computers upon individual and community capacity. A list of these informants, identified either by their real name or by a pseudonym, is provided in the 'Cast of Characters' at the end of this chapter. A graduate student (Maureen Kearley) and a local resident (Keith Davis) conducted the interviews, summarizing them (with permission) as a collection of personal stories (Davis and Kearley, 2005). In the case of Brother Jim, and where information is taken from personal stories approved by the subject, the full name is used. Where information is from full interview transcripts, or from personal stories where the person chose to remain anonymous, a pseudonym is used. Further rounds of analysis of full interview transcripts, using NVivo software (Bella and Kearley, 2005), also informed this account of 'Geeks who Care'.

Gender and caring

Caring theory originates in Carol Gilligan's (1982) critique of Kohlberg's work on the development of ethical reasoning in children. Kohlberg described boys as developing through a number of stages towards a facility in abstract reasoning about right and wrong (1981, 1984). Kohlberg considered this facility as a sign of mature ethical reasoning and found that girls favoured relational ethical reasoning, considering the impact of a decision on those they cared about rather than more abstract considerations of 'right' and 'wrong'. He identified this as less mature and even deficient moral reasoning, in comparison with the boys he studied. Feminist scholars, alert to the sexism apparent in his research, argued that the application of measures based on the study of boys to the study of girls exemplified androcentric bias (Vickers, 1984). Gilligan (1982), beginning instead with the study of girls and women, showed women-based decisions about such matters as abortion, on the effects of their decisions on those they cared about—a form of ethical reasoning now identified as an 'ethic of care'. While the works of both Kohlberg and Gilligan have been revisited, critiqued and re-analysed many times since this debate exploded in the 1980s (Larrabee, 1993), a focus on caring has become a significant feature of the study of gender relations.

Gilligan's work, and responses to it, laid the foundations of 'caring theory and research'. This field includes studies of 'caring' occupations and of the unpaid caring provided in families and communities (Baines, Evans and Neysmith, 1998). Caring occupations include nursing, social work (Bella, 1995), teaching and librarianship, and are prevalent within the health care system. Most caring occupations were traditionally filled by women, and tend to be underpaid relative to occupations where men predominate (Armstrong and Armstrong, 1978). Unpaid caring has been a significant and often invisible feature of most women's lives, and includes childcare within the family, and responsibility for the care of older relatives (Armstrong and Armstrong, 1978; Baines et al., 1998).

This invisible and voluntary caring work is essential to the maintenance of the social networks identified as a key health determinant. In contrast, work related to machinery and technology (such as engineering in all its forms) (Spender, 1995) and to management and administration (Struthers, 1987; White, 1988) has generally been understood as men's work, and more generously compensated. Caring embedded in such work is more visible than in social care occupations.

Caring theory and research has been enriched by Tronto (1993: 105–108), who identified four 'analytically separate, but interconnected, phases'. She identified the first as 'caring about' as 'the recognition in the first place that care is necessary'. The second is 'taking care of', or 'assuming some responsibility for the identified need and determining how to respond to it'. I call this 'executive caring'. The third phase, 'care giving' or 'direct care', involves physical work and direct contact with the person receiving care and is what we traditionally understand as 'care'. According to Tronto, 'the final phase of caring recognizes that the object of care will respond to the care it receives', and is described as 'care-receiving' or 'being cared for' (107–108). Caring work, both voluntary and paid, has relied increasingly on expanding information technologies. Health information is available on the Web to both professional and lay users, and computers are essential to work place recording and control and to the management of community-based organizations such as MacMorran. Referring to this community-based setting, we show how gendered processes of caring interweave with gendered technological work. In order to foreground this discussion, in the next section we draw upon the gender and technology debate.

Gender and technology

Several concepts from the gender and technology literature are useful in studying MacMorran's public access computer programme. The term 'technology' can be understood as including technical knowledge, such as knowledge about how a computer works and how to fix it; and/or a set of activities and practices, such as the use of the technology to complete a task or produce a product; and/or as the equipment itself, in this case a set of public access computer terminals with Internet access (Wajcman, 1991: 14–15). A number of theorists have argued that a gender divide pervades technology in general and computer technology in particular. In terms of knowledge, men tend to predominate in computer classrooms from grade school (Cooper and Weaver, 2003) to post-secondary training programmes (Balka and Smith, 2000: 3). The IT workforce is also dominated by men (Cooper and Weaver, 2003).

As a result, computer systems design is dominated by men, and the products of their work are imbued with traditionally masculine values such as speed, strength and competitive success, rather than more traditionally feminine values of intuition, tenacity and compassion (Wajcman, 1991: 18). This dissonance between traditionally feminine values and those embedded in software have led many women (and some men) to experience 'computer anxiety', or 'feelings of discomfort, stress or anxiety' in response to computers (Wajcman: 13–16). Linked research also suggests that males and females appear to approach computers differently. Boys tend to learn about the technology by playing with it while girls appear to want to understand how it works before they begin to play (Spender, 1995). Men tend to see playing with the computer as an end in itself, and the machine as having intrinsic value, even to the point of displaying addictive behaviour patterns of computer use. For women, the machine is more of an instrument used for a purpose, which they would fulfil without a computer if that were possible (Woodfield, 2000: 89–120). I use the terms 'intrinsic' and 'instrumental' to identify the two poles of this continuum in terms of approaches to computer technology.

Traditionally masculine and feminine representations of technology and related attitudes towards its use and value, including such concepts as 'computer anxiety', and 'intrinsic' or 'instrumental' interest help us understand the ways in which gendered patterns of caring and technology interweave at MacMorran. As the stories of men and women unfold in this chapter, we show how both genders learned about the instrumental value of computer technology. In this context, a small group of young men and one woman with intrinsic interest in the technology were inspired by the Centre's ethic of care to do both the executive caring and direct caring work needed to disperse computer literacy throughout their community. With respect and affection we coined the phrase 'Geeks who Care' to describe this group.

The Geek who Cares

The typical computer support person is probably male, with an intrinsic interest in the computer, and, according to Spender (1995), is likely to treat women's computer difficulties with ill-disguised contempt. She contrasts this unfortunate reality familiar to many of us as we have tried to master new equipment and software, with the ideal computer support person who (although male) is sympathetic and respectful:

> Rarely is the computer support person like the thoughtful young man who comes to get rid of bugs in my system. He respects my work and sees his role as helping me to be able to do it with the aid of a computer. He makes no attempt to show me up as ignorant but designs programs that make it easier for me to do what I want to. He stands by while I stay seated at the keyboard, and he carefully talks me through the steps until I am confident that I can do it myself when he leaves.
>
> From accounts I have heard, this is an unusual experience. Much more common is the computer support person who casts the client aside and after a series of flamboyant clicks which emphasize his expertise and mystique, declares that all has been fixed, it was merely a matter of doing 'X Y Z'. He proceeds to depart—looking bored and disdainful.
>
> The woman who asked for support and instruction is none the wiser. It would be understandable if she were less confident. If she were left feeling frustrated, stupid—and that she had been a bit of a nuisance bothering the big man.
>
> (Spender, 1995: 174)

We would identify the young man who fixes Spender's computer with such care and respect as a 'Geek who Cares'. The stereotype she uses in contrast is all too familiar to women who have sought help with computers. Our research at MacMorran shows that although men may be among the first to use and become familiar with the technology in the Centre, in the context of a pervasive ethic of care, some assumed the role of 'Geeks who Care', engaging in both executive and direct caring for both male and female computer users, even those afflicted by computer anxiety.

Brother Jim and the origins of MacMorran's ethic of care

The Centre's tradition of caring originated with the work of Brother Jim McSheffrey, a Jesuit living in the housing development. His leadership in direct caring work directly influenced many men and women, as he reached out to new comers, giving concrete help, drawing out the lonely and comforting those in pain. These accounts show how he helped:

I didn't move into the house, my husband had to move into the house. I had nothing. Brother Jim came knocking on the door to see if we needed anything. I remember Bill calling and saying 'there's a man at the door, a brother or something, to see if we need anything'. I remember laughing and saying 'that's Brother Jim McSheffrey, he does a lot for people'. I actually didn't have a bed or a kitchen table set. By that night, Brother Jim had placed a bed and a kitchen table set in my house (Donna).

I remember that too as my son had an accident and split his lip and Brother Jim had taken us to the Janeway (the children's hospital) (Keith Davis).

I used to play hockey with Brother Jim then. I was about fifteen, sixteen. Fifteen or sixteen years ago playing hockey with Brother Jim who used to take all the boys out to Buckmasters Circle to play hockey at the Rec Centre (Dave).

However, Brother Jim was not technologically adept, and needed his neighbour's help. As Chuck commented,

I even showed the late Brother Jim how to … I forget what it's called right now. I showed him Merge, merge letter, that's it! I think it was seventy odd letters he sent around the world looking for other McSheffreys.

Brother Jim died in 1999. An enthusiastic berry picker (a favourite activity in Newfoundland in the later summer), he would distribute bags of frozen blueberries to everyone in the community. Unfortunately, he had a habit of working backwards when picking through a berry patch, and although the entire MacMorran community was very shocked and desperately grieved to hear that he had fallen to his death over a cliff, no one was entirely surprised. Since Brother Jim's death, the Centre has moved into a new million-dollar building with five times the space. Brother Jim is acknowledged in a stained glass window, from which he waves cheerfully over the gymnasium. His legacy is acknowledged:

Brother Jim was certainly very important in that whole process. He nurtured a lot of people and he nurtured the community. And we really miss him since his demise, his very untimely demise in 1999. (Jack)

Brother Jim and men who care at MacMorran: Mike, Keith, Brian and John

Mike Wadden, Executive Director of the Centre (an executive caring position), started as a summer counsellor (a direct care role), but took over when the previous manager was sick. Mike held this position throughout the development of the Centre's public access computer programme. His role was primarily one of executive caring, trying to overcome the disorganization he had inherited and to foster mutual respect (that is, caring about). However, Mike did not consider himself a 'techie' and, like Brother Jim, relied on others' instruction. When the first computer arrived, Mike was sent off to learn Windows and Word Perfect, and another executive director taught him accounting software. Over time, familiarity with email and the Web became essential, so anyone working at the Centre needed to use computers. As Mike said, 'At the end of the day the expectation was that everybody could function with the computer', and everyone would have to learn.

Brother Jim had helped Keith Davis as a neighbour and friend struggling to raise four children on his own. Keith says his volunteer work with the Centre did not include direct caring (he did enough of that at home), but he served in many executive caring roles. Brother Jim took him to his first Tenants' Association meeting in 1989, and subsequently he served on the Executive for 11 years, with nine as President. He also spent ten years on the Centre board, with three as President. Because of this history of commitment, Keith was hired as Building Project Facilitator for the new Centre. Mike Wadden expected him to use computers on the job, but Keith 'wouldn't go near them' due to classic 'computer anxiety'. Instead, he brought his typewriter to work. Hearing the clatter, Mike said, 'That goes! You're getting your computer.' So Keith got an appointment with the CAP intern, and 'that's how Mike got me started'. For a while he resisted email, but Mike 'caught on' and Keith now uses email for committee work and to communicate with friends and relatives. He used the computer to work on this ACTION for Health project, but still does not entirely trust the new technologies.

Brian Conway was recruited by Brother Jim for the Tenants' Association in 1993. He has served in many executive caring roles. He wrote minutes by hand, and then with a typewriter. After the Centre got a computer he 'got bold one day and decided I was going to use it' (suggesting initial computer anxiety). He learned a lot on his own, 'sitting down and figuring stuff', suggesting a more masculine approach to learning about technology, but if he needed to know anything about what he was doing he would 'ask my son Dale, or Donny Howell, and they would show me or tell me what to do'. Brian can now use Word Perfect, Print Shop and email and searches the Web, both at the Centre and at home. His approach to the technology is largely instrumental.

Brother Jim also recruited John for direct caring work, asking him to help collect for and deliver Christmas hampers. John continues to do this and was proud to be the first man (apart from Brother Jim) to volunteer for the local food bank. In addition to this direct caring work, John also became involved in executive caring on the Tenants' Association and the Centre's board. John learned computer skills in school, and his direct caring extends to sharing his computer skills. He taught various skills to Brother Jim, Mike Wadden and to Keith Davis. He was the first of the men involved in the Centre to develop sufficient computer knowledge to repair computers and programs and to teach others how to use them, although his interest remained instrumental rather than intrinsic.

The commitment to caring demonstrated by these four men suggests that an alternative form of masculinity developed under Brother Jim's leadership, not concerned with technology and technological expertise, but with caring and community. As Dale Conway said,

> Brother Jim was like that, he used to go out and get people 'on the go.' I think it was his vocation to make the community a better place to live in, probably part of his being a Jesuit brother.

As we demonstrate below, some younger adults, including Dale, developed an intrinsic interest in the technology. This combined with their commitment to caring, led to them becoming Geeks who Care.

Brother Jim and women who care at MacMorran: Denise and Bernice

Brother Jim encouraged Denise Bradley to involve herself and her children at the Centre, first as a volunteer, then with the Tenants' Association. She 'got more and more involved'. She had learned computer skills at a private career college, but had not used them. Mike Wadden asked her to help with the books, and then hired her to teach 'basic computers to people who didn't even know how to turn on a computer', with Donny Howell. She gives credit to people like Mike and Brother Jim for believing in her when she 'was a mess' and did not believe in herself. When she was offered a paid job, she was not sure she could do it, but Mike said she was ready and she held that job for three years. Personal support and encouragement appear to lead to employment, and an increased likelihood of improved health. Denise is now anticipating a career in recreation leadership: she is familiar with computers, but really wants to do caring work.

Bernice participated in Centre Christmas parties when her son was little, but her later involvement in executive caring roles encouraged her to acquire computer skills. Brighter Futures[3] sponsored computer workshops, and Mike Wadden helped her do their minutes on the CAP computers. Brother Jim persuaded her to join the Tenants' Association, and she also joined the MacMorran Board. Like Denise, her approach to the computers is instrumental rather than intrinsic. She claims that she still needs more help, suggesting some residual computer anxiety, because of difficulty in saving files and with the Internet, but says the Centre's computer classes are inconvenient for her.

The men and women discussed above come from a generation introduced to computers as adults. They were all profoundly influenced by Brother Jim's encouragement and caring, and inherited his commitment to caring and community. For them, the ethic of care is the primary commitment. The technology is merely an instrument to enhance their executive (and in some cases direct) caring work. In this case, the more traditional masculine and feminine roles and identities related to both caring and to technology merged, as men learned to care and women learned to use computers to expedite their caring. This parallels Simpson et al.'s conclusion (Chapter 2, this book) that stereotypical portrayals of health caring work as either male or female are increasingly inappropriate.

Geeks who Care: Dale, Donny, Donnie and Angela[4]

Most of the men and women described above have teenage or adult children. Brother Jim influenced the children as well as their parents, taking boys to recreation activities in the city. Like their parents, they developed a commitment to caring for the community. However, unlike their parents, they became computer literate as children, learning both at school and from MacMorran. Several developed an intrinsic interest in the technology, combining that with a commitment to caring, becoming Geeks who Care. Most were young men, but Angela also qualified both as a Geek, and as one who cares.

Dale Conway moved into the community with his parents. After Brother Jim got his father (Brian) into the Tenants' Association, his father brought him along too. The computer activity at MacMorran was a major influence on Dale's life, and he has played a major role in updating the hardware and supporting others to computer literacy. First asked to type up minutes for the secretary, he stayed to 'play' with the computers—'fixing and breaking and fixing again'. This intrinsic approach to computers appeared almost addictive at times, and would infuriate other users, until he learned to share his knowledge about computers with staff and volunteers. Whilst still in high school he was given the job of installing all the new CAP computers, even though his skills were largely self-taught and acquired through his intrinsic interest in the machines:

> The past 12 years have been speckled with pivotal points in my life and MacMorran, and having access to computers, has been one of them.

Dale described his work with the Centre's computers as including letter and report writing, keeping attendance logs, informing important officials and Centre supporters on an email list and installing security to prevent access to pornography. He fixed things, 'troubleshooting, fixing the printer, finding out what network problems we were having'. He created 'a little booklet: The Four or Five Things You need to Know With the Computer'. This work was typical of the 'executive caring' more characteristic of men's caring work. However, Dale and Donny both helped children avoid the dangers of Internet predators, and many Centre participants described getting help from Dale, which suggests he was more involved in direct caring than he confided in our interview: he was becoming a Geek who Cares. From his point of view, if he had not moved to this community, he probably would not be running his computer business today.

> Being a part of a community, like Brophy Place (a street near MacMorran), can provide you with a sense of security because when you're there or going around the community—everybody knows your name. This is a good thing. It's like my work with computers. I may have started at MacMorran but my name has gone further. I've never had to apply for a job or do a resume. People have come looking for me or I'm recommended. It all started with having the opportunity to get experience. This happened for me at MacMorran. Having experience is even more important than having the skills, it's the experience that counts because it hones the skills.

Donald D. Howell (Donny) was one of the children that Brother Jim drove to recreation programmes, and he became a MacMorran volunteer as a teenager. He performed both direct and executive caring, helping with homework, floor hockey, special events, the newsletter, fund raising, getting contracts, making contact lists and writing proposals. Although initially introduced to computers in school and at a private college, Donny learned more from Dale about hardware, software and troubleshooting, later developing a more intrinsic interest in the technology. He and Dale (and later Denise) taught the Centre's earlier computer classes. Donny qualifies as the Centre's second Geek who Cares. When the Centre moved to its new building, he and Dale networked the server with the computers and printers within a week.

Although Donny describes most of this work in terms of executive caring, he also acknowledged doing some hands-on caring. His own description of helping his sister is more typical of his role as a Geek who Cares (i.e. responding to her personal interests rather than following a set of modules).

> My sister knew nothing about computers. At first she was not very interested but when I showed her what she could do she started to learn. Now she researches astrology which is her favourite hobby and she writes poems and stories. Some of them have been published in the newsletter.

When teenagers began to use chat rooms, the Centre's board became concerned that sexual perpetrators would use them to recruit children, Dale's response to these concerns highlights another direct care role. He and Donny found that some teenagers had been giving out the Centre's telephone number, putting themselves and everyone else at risk. He then talked to them through Internet security procedures including never disclosing their location or any other personal information on the chat line. The Board eventually disabled the chat lines altogether. During his interview, Donny emphasized his more visible executive caring work, although he admits to doing direct caring work. In the next section we show how this work is acknowledged and valued by those he helped.

Donny's half brother, Donald R. Howell (or Donnie) had played hockey with Brother Jim. When he moved into the community with his family four years ago, he quickly volunteered to contribute computer knowledge acquired at college. He helped solve computer problems, and began to teach computing at the Centre. This helped refresh his computer knowledge and improve his keyboard skills. He uses computers all the time, and got his present job through the Internet. Donnie described helping Keith Davis:

> Poor Keith was told listen you got to set up an e-mail and Keith was like 'What! Set up an e-mail! I can't be doing that!' The he came to me and I said, 'yes I'll help you set it up. I'll show you how to do it.'

Donnie's intrinsic interest in the technology combined with his apparent commitment to caring qualifies him as the third Geek who Cares at MacMorran.

Angela McIntyre is our first female Geek who Cares. New to Mac-Morran, she soon realized a visit to the city library required a bus ride but that the computers at the Centre were only a 2-minute walk away. Angela used Centre computers to prepare resumes and check her email. She left the province for school (including computer training), and on her return was recruited as instructor for the literacy programme 'Computers Plus'. She describes her real love for computers and she not even wants to get paid, which confirms her intrinsic interest in the technology. Angela describes the work as 'a wonderful experience, seeing computers contribute to literacy'.

> One of the people I taught computers to just learned how to read last year. Him learning how to use a computer on top of that—was phenomenal for him. He actually wants to go back to school now. His wife walked up to me and thanked me and said 'you have changed his life'. It's freaky because I'm not used to that. So to feel that you've inspired someone is phenomenal—oh my god!

To Angela, her direct caring work was the most visible and significant part of her job, contrasting with the male Geeks who Care's emphasis on executive caring. Because she combines a traditionally feminine skill of direct care with the more masculine intrinsic interest in technology we include Angela as a Geek. However, direct care is for her more visible and valued than it is for our male Geeks.

Getting help from Geeks who Care[5]

The data from MacMorran provides a pertinent example of different but interconnected perceptions of and attitudes towards technology. Although they acknowledged giving individual help, in their interviews

the male Geeks who Care (Dale, Donny and Donnie) emphasized technological problems, policies and teaching materials. Direct help was
mentioned, but not as the central task. In contrast, Angela, our female
Geek, highlighted the significance of her direct caring work. This leads
us to ask the following questions: Were they actually doing the same
work, but giving it different emphasis in their interviews? Or was
Angela's help qualitatively different from that given by Dale, Donny
and Donnie? To more fully understand the gendered dialogues from our
Geeks, we analysed data from the MacMorran residents they helped.

In the early days of the computer access programme most of the computer literacy support work was done by Dale and Donny. Dale's father,
Brian, learned a lot himself, but if he got stuck he would 'ask my son
Dale, or Donny Howell, and they would show me or tell me what to
do'. Kimberley would ask Dale or Donny, 'how does this work?', but
also learned some of it on her own. As Stephany told us:

> When I used to come down when they had people that used to show
> us how to get into programs. Like Donny Howell, he used to help me
> get into a lot of programs and help me a lot of the time.

Steven said that Donny 'actually taught him 90%' of what he knows
about computers, 'and that was at the Centre'. And as confirmed by
Patricia, Donny was often more available than staff because he was a volunteer, and 'would take you aside and help you with what you needed
to be done'. Paid staff faced deadlines and could be less available:

> Finding a time … was hard, and then having to ask someone who is
> busy themselves, to remove themselves from their computer and let
> you sit down. You feels right guilty.

Laura describes how Donny and Dale responded to her need for
individualized and 'direct care'. If she got stuck she would say,

> 'Oh! I don't know how to do this. Can someone help me do it?'
> Donny Howell was here, and he was good with computers, and
> Dale was here and he was good with computers. When I needed
> help they'd help me. They'd teach me, things in Word Perfect or
> Microsoft word.

For her they were Geeks who Cared. She now works in another community centre, partly because the positive feedback at MacMorran helped

her gain confidence and computer skills; however, she still returns to MacMorran for community events.

During the field-work phase of this research, Angela was paid to offer computer instruction and support and was the most available. Those whom Angela taught found her classes 'really good', because even those who were computer literate learned something new, such as 'little bit more about Word Perfect'. In Eddy's words, 'Angela's pretty good, she knows lots of stuff.' As a CAP intern, Angela also did individual tutoring, which included individual work with people with no computer knowledge:

> So Angie, the one that works here, she started some stuff with me before Christmas. And like basic things like how to turn the computer on, and this is the mouse, and do you know what hardware is and software. I didn't know what it was. I didn't know. And tabs, backspace, how to capitalize something.

To this inexperienced computer user the group classes were overwhelming:

> I was sat in there when people were running right through things and I'd be like, 'Oh my god I put an "A" in there where I shouldn't have. How do I go back?!'. I didn't know. So it may sound pretty minor but it's really important to me, right. She's a good person (Angie) and she sits down with me. I'm really starting to, you know, get into it.

Angela helped anyone stuck with a computer problem:

> Angela, who's our CAP intern, she's the one does our computer teaching and she does one on one tutoring. So whenever we have a problem, or I have a problem, I call Angela and I say 'ah ... I'm doing something and I can't get this to come up'.

Freida, a programme leader with an instrumental interest in the technology and a commitment to caring, could also turn to Angela:

> So whenever we have a problem, or I have a problem, I call Angela and I say 'ah ... I'm doing something and I can't get this to come up. How do I do it?'

Those receiving computer support from our Geeks who Care describe similar experiences of being helped, from both male and female Geeks, and also (although not reported here) from other staff, residents and volunteers with a more instrumental approach to the technology. This is consistent with other accounts of receiving help at MacMorran in areas unrelated to technology—such as Brother Jim's outreach to newcomers and John's delivery of Christmas hampers. The ethic of care at MacMorran permeates all centre activities, and is understood as a guiding principle for the Centre's management and development. In large part because of the heritage from Brother Jim's influence, caring at MacMorran transcends gender—both men and women are involved in both direct caring and executive caring work. This is true whether or not that caring work involves technology. Geeks who Care express this commitment to caring in the ways that they help others to develop computer skills and solve computer problems.

Gender, caring and Geeks who Care

This story began with the quite remarkable Brother Jim. He understood the significance of direct caring, engaging in it effectively himself. His charismatic leadership led many men and women into both executive and direct caring roles in their community. We have shown how his legacy encouraged men to become engaged in caring work, both executive caring and the direct caring that is not stereotypically a masculine activity. This chapter has argued that this formed the basis of an alternative masculinity at MacMorran, creating the foundation for development of a computer programme in which caring was central.

The community access programme (CAP) at MacMorran introduced a generation of mature adults to computer technologies, enabling their use in various executive and direct caring roles, as well as in personal interest pursuits. This strengthened the quality of programming at Mac-Morran, linking it to all five of the key determinants of health: income, education/literacy, employment, involvement in social networks, and the strength of those networks. Enhanced computer literacy enabled people to feel more confident, to find jobs, to advance in school, to develop careers and to work together to enhance activities in the Centre and the surrounding community. For both men and women, however, the interest in technology remained instrumental, useful in supporting other activities such as care work, but not of intrinsic value.

A second generation, who had participated as children in recreation programmes with Brother Jim, were also influenced by his legacy. However, unlike their parents, they were introduced as children to computers, both at school and at MacMorran. Some younger adults retained an instrumental interest in the technology, characteristic of their parents' approach and used it to pursue careers in caring. On the other hand, we identified four with an intrinsic interest in the technology. Influenced by Brother Jim's legacy, they became what we have called 'Geeks who Care'. This group includes three men and one woman. Our Geeks are described enthusiastically by those who received their help and appear to emulate Spender's perfect computer support person. Although, as we have shown, caring work transcends gender expectations at MacMorran, we found a difference in the extent to which our four Geeks acknowledge the significance of their own direct caring work. This appears to be a crucial site of gender difference. Although those who were helped described the ways in which they were helped in very similar ways, the Geeks differed in their emphasis on this aspect of their work. Dale did not tell us much about the direct caring aspect of his work, and Donnie and Donny told us only a little more, but we can 'read' the gender dimensions through their dialogue. Unlike her male counterparts, Angela emphasized the direct caring part of her work as the most valuable and rewarding. Our data about Geeks who Care at MacMorran suggests that direct caring work for the men is a sub-text, but it is central to the narratives of the female Geeks. This is consistent with other caring research which argues that direct care is 'women's work' that is often devalued and rendered invisible (Baines et al., 1998). This suggests that even when men do direct caring work, they may not experience it in the ways that women do. Even when men do direct caring work, they may not value that work enough to describe it to others. This further contributes to the invisibility of the direct caring work of both men and women, and to the persistent assumptions that devalue such work as 'women's work'.

MacMorran Community Centre Cast of Characters

Brother Jim McSheffery	Jesuit brother, original founder of MacMorran, engaged in direct caring, acquired computer skills as a mature adult.
Mike Wadden	Executive Director of the Centre for much of the field-work period, Mike supported development of the CAP site and of spreading computer literacy.
Keith Davis	Long-term resident and friend of Brother Jim's, who raised four children alone. A community leader, Keith overcame his computer anxiety when the Centre hired him to manage their new building project.

Brian Conway	Recruited by Brother Jim. A community leader who learned his computer skills from his son via the Centre's computers.
John	Recruited by Brother Jim for both direct care and local leadership, John (a pseudonym) was the first resident to begin teaching others to use computers.
Chuck	A pseudonym for an adult male who acquired his computer skills at the Centre, with support from his son.
Jack	A pseudonym for a long-term resident and supporter of Brother Jim.
Denise Bradley	A married woman with children and a leader at the Centre, Denise gave Mike and Jim credit for encouraging her to get back on track and involved in caring work.
Bernice	A pseudonym for a woman who overcame anxiety, learning to use the computer to do secretarial work for the Centre.
Donna	A pseudonym for married woman with two teenage daughters.
Marg	A pseudonym for a single parent raising her son at MacMorran.
The Next Generation:	
Dale Conway	The first real 'Geek' at MacMorran, with a clearly intrinsic interest in technology. This combined with his commitment to caring created a career as a Geek who Cares.
Donny Howell	Donny is the second Geek who Cares at MacMorran. He was influenced by Brother Jim at a young age and later worked extensively as both volunteer and staff with the Centre's programmes.
Donald R Howell (Donnie)	Donnie, Donny's half brother, was also involved in Brother Jim's recreation programming. He moved into the community later than others, helping his brother with the Centre's computers. He qualified as the third 'Geek who Cares'.

Angela McIntyre   The Centre's first female 'Geek who Cares'.
Those helped by Geeks who Care included:
Kimberly, Patricia, Steven, Laura, Eddy, Freida and Stephany
(pseudonyms for residents, staff and volunteers).

Notes

1. St John's is the capital city of Canada's poorest province, and is located on the eastern coast of the island of Newfoundland known affectionately by Newfoundlanders as 'The Rock'. The province joined the Canadian Confederation in 1949, and has since been devastated by the collapse of the Atlantic cod fishery.
2. Ethical approval granted by the Interdisciplinary Committee for Ethics in Human Research (ICEHR) at Memorial University of Newfoundland.
3. Brighter Futures is a government-sponsored programme to improve the life chances of children born into low-income families, and emphasizes nutrition, parent education and peer support.
4. Some in this younger generation retained the instrumental interest in technology, and therefore are not included as Geeks who Care. Gary, for example, is more committed to a career in caring than to the computer technology that helped him get there.
5. The Geeks who Care were not the only ones who help with computer problems at MacMorran. With a generalized ethic of care, anyone with skills was expected to help anyone else. Mike, Michelle, Sonya and Denise, and many others, all with a commitment to caring and an instrumental interest in computers, were also acknowledged for their help. However, in this section we focus on help received from our Geeks.

4

Cyber-Burdens: Emerging Imperatives in Women's Unpaid Care Work

Roma Harris

As mothers, daughters, wives, community volunteers, and in countless paid work roles, whether they are teachers, nurses, hairdressers, librarians or lawyers, women are called upon to react to and act upon the health concerns of others. As carers and care-givers women are expected to demonstrate empathy, concern, and interest in the ailments and health issues that affect children, partners, friends, and colleagues, as well as workplace customers and clients. Along with lending a sympathetic ear, women often serve as important intermediaries of health information, even though they may have no recognized links or responsibilities to the formal health care system. By finding and passing along information about illnesses, wellness strategies, and treatment options, and by locating services and resources, women play pivotal roles of support and harm-reduction in their communities. This job of care and connecting is largely invisible and, even if undertaken in the context of paid work, is usually under-appreciated and un- or under-compensated.

The health policies of Western governments are changing to reflect an increasing emphasis on the role of individual citizens in maintaining and managing their own health. 'The notion that the state should care for the health of its citizens, long seen as a fundamental principle of welfare states, is increasingly replaced by the expectation that citizens should play a more active role in caring for themselves as "clients" or "consumers"' (Henderson and Petersen, 2002: 1). In Canada, for example, a group that argued for the development of a national e-health strategy wrote:

> Our health system is being fundamentally transformed. We are moving from a focus on acute care and cure to a broader vision that

includes health promotion and disease prevention: from a focus on central control of institutions to regional support of home and self-managed care; and from a reliance on medical specialists to *a recognition by all citizens of the need to assume greater responsibility for their own health.*

(Canadian Network for the Advancement of Research,
Industry and Education, 1997: iv, emphasis mine)

Accompanying this shift in responsibility is an increase in women's workload since women often assume, by proxy, many 'self' care duties on behalf of family members and others in their social networks. A significant component of the work associated with these duties involves searching for information relevant to decisions about treating and managing illnesses, as well as keeping abreast of new information, particularly with regard to activities that may affect health outcomes, such as food selection.

To support what Bunton and Crawshaw (2002: 187) describe as 'enterprising, self-managing citizenships', governments encourage citizens to use e-health services, such as the United Kingdom's NHS Direct Online, to enable the information-seeking such self-management entails. Whether or not these government-sponsored web services are actually used by members of the lay public for whom they are intended, there is considerable evidence that the Internet is emerging as an important enabler of health information seeking for 'connected' citizens, especially women (see, for example, Fallows, 2005). Ironically, because of its power, convenience, and the vast range of information it makes accessible, the Internet itself may be creating imperatives that not only add to the burden of women's health care work, but accelerate the pace at which they are expected to do it. In this chapter, I describe the pressures women face to keep themselves informed about health and discuss how even the most motivated health information seekers are sometimes daunted by the challenges they encounter. These challenges, all of which are aspects of what I call 'health info(r)mediation', include, among others things, finding relevant information, deciding what information is credible, interpreting complex data, managing the emotional fallout of 'bad news', and applying conflicting or competing advice.

Women, health, and care work

Caring may not be good for women's health. According to Denton, Prus and Walters (2004: 2592), 'the everyday stressors associated with women's social roles' may have a negative impact on women's health. One such role involves responsibility for family health, an activity that is more highly associated with chronic stress for women than it is for men (Denton et al., 2004). In spite of a large body of evidence indicating that people who have close connections to others are likely to have better health than those who are isolated, the relationship between social networks and health is less straightforward for women than it is for men. For example, among women over age 50, being socially connected may be associated with *higher* rates of illness and death, likely because of the type and degree of responsibility embedded in their relationships with others in their personal networks (Barefoot et al., 2005). As Shumaker and Hill (1991: 107) explain, women 'cast a "wider net of concern", have a greater inclination to get involved with care giving activities, and are more responsive to the life events of others' and they are 'more likely to be involved in all forms of help giving than men', such as looking after elderly parents.

A significant aspect of domestic care-giving involves locating and keeping track of health-related information. Women often assume the major responsibility for this work on behalf of family and friends (see, for example, Warner and Procaccino, 2003). In fact, Moen and Brennan (2005) argue that this home 'health information management' role is becoming so significant that it is now 'an essential responsibility' (648), particularly for women who comprise the majority of 'self-identified health information managers' (651). The expectation that women will take on this work is not only internalized by women themselves, but also reinforced by others in their social networks. For instance, focus groups conducted as part of a suicide prevention study revealed that many Australian men expect their wives and girlfriends to look for health information and negotiate health-related concerns on their behalf (Peta Wellstead, personal communication, July 2006). Similar expectations are reinforced in media representations of women's domestic roles, in public health promotion campaigns, and in the marketing of health-related products where there is persistent messaging about women's responsibility for family health. With respect to diet and nutrition, for example, women are portrayed as 'nutritional gatekeepers' who are responsible for their families' eating habits, especially those of children, and are seen as the culpable parties if their family members fail to eat 'properly' or are overweight. As one popular health web site proclaims, 'the more a mother knows about nutrition, the less likely her children are to be overweight' (WebMD University).

In addition to women's roles as family health information seekers and health knowledge gatekeepers, they also serve as lay consultants who pass along advice about symptoms, treatment strategies, and whether to seek professional help. Stoller (1993) describes as 'lay care' women's efforts to 'provide instrumental assistance to people experiencing illness and disability, as well as information about health promotion, symptom interpretation, and response, and management of disease' (151). Although Stoller frames lay care as the 'informal, unpaid care provided by family, friends or neighbors' (151), there are also many occupational roles in which women perform similar caring tasks that extend beyond the jobs for which they are actually paid, such as nurses who make referrals for seniors to local trades and services while conducting community footcare clinics (Pettigrew, 2000) or hair stylists who soothe and counsel emotionally distraught clients (Lawson, 1999). This extra work of care is usually done outside of, or in spite of, profitability considerations and practice guidelines.

Whether managing their own health or supporting the health of others, women often rely on other women in their friendship, family, and community networks to help locate the information they need to cope with immediate health-related situations and to keep up to date with information that might be useful in the future. In interviews with women living in a medically underserved rural county in Canada, we learned a great deal about the efforts women make to find and apply health information (Harris and Wathen, 2007). One woman explained, for instance, that for her, the best source of health information is: 'The informal system that's around you of colleagues, friends, family, resources. I guess I have a bit of a bias where I feel put off by the formal system because it's not user-friendly, it tends to be bureaucratic, and you have to work with it to get what you want.'

Being aware of and having access to a wide range of information sources and alternative options is important to women's roles as family health gatekeepers. As Wuest (2000) explains, before women will negotiate with professional helpers to get what they need for themselves and others, they have typically experienced adversity and disillusionment with 'systems that failed to help, provided inadequate help, or made things worse' (p. 57). Contrary to a commonly held view that women overuse health care systems, they are sometimes more likely than men to self-treat and limit their use of services, particularly in response to social expectations that they should be stoic and uncomplaining in the face of their own illness (see, for example, Vlassoff and Moreno, 2002;

Harris and Wathen, 2007). Meadows, Thurston, and Berenson (2001) observed that as a result of 'messages about busy doctors and illegitimate visits' (460), women carefully regulate their use of the formal system and avoid using it for 'merely preventive purposes' (460) and turn instead to friends, family, and others in their local community networks. According to Wuest, the negotiation process women undertake to deal with their own and others' health concerns involves 'reframing responsibility, becoming an expert, harnessing resources, and taking on more' (58). Reframing occurs when women realize they must assume an active role if they are going to succeed in 'making resources work' (Wuest, 2000: 59). Becoming an expert obliges women to find and apply new knowledge to their situations. To acquire this knowledge, they are more likely than men to rely on 'alternative' providers (Meadows et al., 2001), consult self-care books (Hibbard, Greenlick, Jimison, Kunkel, and Tusler, 1999), and they use the Internet 'to supplement health information from traditional sources' (Pandey, Hart, and Tiwary, 2003: 179).

Health help-seeking and the Internet

Health information seeking via the Internet is becoming increasingly common. At least a third of Europeans (Eaton, 2002) and 80 per cent of adult Internet users in the United States (Fox, 2006) have searched online for information about health topics. However, whether they are professional health service providers or lay users, the majority of online information seekers use less than optimal search strategies, such as brief sessions during which they enter a few terms into general web search engines. Such techniques often produce 'misleading or unrelated information' (Spink et al., 2004: 45). Lay searchers may have difficulty judging the reliability of health information they retrieve and/or be unaware of the need to assess the potential bias of sources that produce or transmit the information they find, and few actually check the source or dates of information located online (Fox, 2006). Information intended for lay use may not 'fulfil the quality criteria required for unbiased evidence-based patient information' (Mühlhauser and Berger, 2000: 824) and, if lay information seekers use natural language expressions rather than scientific or medical terms when searching for health information online, they may retrieve little of the relevant material available (see for example, Sievert, Patrick, and Reid, 2001). In fact, given the underlying design of commercial web search engines (which are built on natural language expressions), Spink et al. (2004: 49) question if it is possible that they can be used reliably to locate credible health information 'when the use of precise clinical terms, descriptions, or concepts is called for'. To overcome such problems, governments and groups concerned with specific disease conditions, such as the American Cancer Society, produce web sites and web portals to link searchers to sources they deem to be reliable and credible. However, in spite of considerable investment and promotion in these services, lay users may not know of, find, or use them, especially the web-based health tools produced through e-government initiatives (see, for example, Hirji, 2004). In a telephone survey of rural residents, my colleagues and I found that 75 per cent of respondents had looked for health information in the previous year and, of these, two-thirds had searched online. However, most relied on keyword searches, only 20 per cent intentionally visited any specific web sites, and few were aware of or had ever used government-sponsored health web portals (Harris, Wathen, and Fear, 2006).

In spite of the quality problems associated with online health information and the challenge of steering citizens to recommended locations on the Web, the ease and immediacy by which health information can be located via the Internet is loosening the strictures of traditional, formal medicine. Lay citizens can, depending on their inclinations and skills, glimpse, peruse, or ponder the evidence (or lack thereof) that underpins health systems and medical practice. This unleashing of information formerly available only to medical insiders has altered the social relations of power and control within the health sector and between that sector and others who seek to use or change it. As Nettleton and Burrows (2003: 179) observed, 'medical knowledge is no longer exclusive to the medical school and the medical text; it has "escaped" into the networks of contemporary infoscapes where it can be accessed, assessed and reappropriated'. Not everyone, however, wants to appropriate medical knowledge or take on the responsibility of finding their own information to make decisions about treatment options and disease management. Many people still prefer to leave this job to their physicians. In fact, in one US survey, 'approximately half of respondents preferred to rely on physicians for information about their condition and half preferred to leave final treatment decisions up to the doctor' (Levinson et al., 2005: 533). Similarly, in a study of middle-aged women in the UK, Henwood et al. (2003) found that more than half had not looked for information prior to seeing a health practitioner, either because they regarded it as 'a doctor's job to inform patients about their health' (597) or because they feared 'being seen to challenge the doctor'

(598). The rural women we interviewed express similar views. As one woman explained, 'I have to be careful not to read too much because I really do trust my doctors.'

Even for those who are willing or enthusiastic health information seekers, searching online can produce results that are overwhelming because of the volume of material retrieved and the complexity of its content. Health searchers' queries may also produce information that is unwelcome, upsetting, and disheartening, and which adds to uncertainty rather than reducing it. As one woman we interviewed said,

> I had a friend do a search on the Internet. That actually scared me even more because you know, when I read it and looked at all the symptoms that I had, it could be this, it could be that, so it could be anything from something very minor to something very major. So, that didn't help. It actually made it worse.

Health info(r)mediating: Invisible, gendered work

The information-seeking work involved in taking responsibility for one's own or family members' health involves much more than simply looking for and locating data relevant to a specific condition or concern. Instead, it means sifting through, interpreting, and dealing with the implications of the information one finds. As a result, whether the information is retrieved from the Internet or elsewhere, domestic health information seeking is best understood as a complex process of exchange in which social (particularly gender) and technical relations are expressed.

When coping with difficult personal problems, information seekers are often looking not only to increase their level of knowledge, but also for emotional support and a relationship with a caring helper/information provider (see, for example, Harris and Dewdney, 1994; Wright, Holcombe, and Salmon, 2004). In fact, the need for a trusting relationship with a person to whom one turns for advice may be just as significant as the actual content of information that is exchanged in a health-related encounter. In a study of female cancer patients' interactions with their doctors, Henman et al. (2002) found that rather than seeking information relevant to treatment decisions, women wanted information that would help them cope with their illness, predict what lies ahead, and assess their physicians' trustworthiness. The women in their study 'partially used the information provided to them to assess their doctor's expertise, and this was important in giving them a sense of trust and confidence that the right decisions were being made' (Henman et al., 2002: 303). Making a similar point, Salander and Henriksson (2005) argue that for people who are coping with serious illnesses, 'communication must be freed from its connection to the conveying of medical information alone and be firmly situated in the patient-physician relationship as a helping relationship' (161).

Unfortunately, the expression of care and relationship that many health information seekers hope to find is not always forthcoming. For instance, physicians may fail to 'provide personalized health information that answers patients' specific questions' (Kivits, 2004: 516). One woman we interviewed explained that in her experience doctors 'don't have a lot of time so you have to be very prepared. You have to go in with your questions and know exactly what information you're looking for. It's almost like you need to know the answers before you ask the questions. I find that a lot of doctors tend to talk at you instead of with you'. This woman's comments illustrate that despite frequent references in contemporary health care discourse to 'partnerships' between health care consumers and providers, the exchange of information during the health consultation is often experienced as a one-way street, in which physicians tell patients about their conditions and treatment options but ignore or refuse information patients may try to introduce into the discussion (see, for example, Rogers and Todd, 2002; Harris, Wathen, and Fear, 2006). If patients anticipate such resistance, it is not surprising that some choose not to mention information they have collected elsewhere to avoid antagonizing their doctors. Others may keep silent because they wish to be perceived as good, compliant patients or because they don't want to waste the doctor's time (see, for example, Henwood et al., 2003; Tuominen, 2004).

When physicians are unavailable, unwilling, or do not have time to engage fully with patients about their concerns, or if health help seekers do not feel entitled to services, are not confident in the formal health system, or are uncomfortable with their doctors' advice, they are likely to look elsewhere for help to locate, interpret, and/or manage health information. In our interviews with rural women, we found that all of them had consulted with physicians about health matters and would do so again in the future. Many described positive encounters with their doctors and most hold physicians in high regard. Nevertheless, nearly all of them had also consulted with and sought advice about health matters from an assortment of other sources, including family members and friends, pharmacists, librarians, health food store employees, massage therapists, various practitioners of complementary and alternative medicine, and even veterinarians. The value the women attached to the help they received, regardless of the source, depended largely on how well the people they had consulted expressed concern or care during an exchange of health-related information. A particularly salient feature of these exchanges was the degree to which helpers were willing to spend enough time to 'really listen' to the woman's concerns (Harris and Wathen, 2007). Similar findings have been reported elsewhere (see for example, Thomlinson et al., 2004).

Beyond the sense of respect 'taking time' conveys to a person who is making an inquiry, complex health information simply cannot be presented, interpreted, and understood without some time passing in a back-and-forth process of exchange, clarification, and assimilation. For example, when applying the results of clinical research to the treatment of individual patients, physicians face a significant challenge in translating the information because of the 'inherent uncertainty' of such evidence (Griffiths, Green, and Tsouroufli, 2005: 1). As Babrow, Kasch, and Ford (1998) explain, 'effective response to illness depends on the ways that patients, loved ones, and health practitioners understand and co-construct the many uncertainties that comprise the illness experience' (3). When physicians do not take the time to achieve such co-constructions or mutual understandings, or they give their time reluctantly, patients and their advocates will turn to others for help and support. Whether in paid or unpaid roles, women are often these 'other' sources of support not only because they may be willing or feel obliged to help, but because social and economic norms dictate that it is not problematic to consume women's time.

Regardless of where it takes place, the health-informing support that women provide to others is work, although it is a form of work that is seldom acknowledged. At home, 'information management, self-care, and health maintenance…remain largely invisible and underarticulated' (Moen and Brennan, 2005: 649). In employment settings, whether needy customers tell their concerns to workers at the hair salon, the health food store, or to waitresses who serve breakfast at the truckstop, the care and health-informing responses they elicit comprise unrecognized work that is provided as an emotional add-on to purchased goods or services. Similarly, when teachers, nurses, family law practitioners, and others employed in female-intensive occupations do the care work involved in health info(r)mediation, they often do so by working outside the boundaries which define their paid roles. For example, nurses who work in tele-health call centres are usually required to follow a standard procedure when responding to callers' inquiries.

Using computerized decision support software, nurses enter terms that describe the callers' symptoms after which they are prompted to ask a series of questions. After entering the callers' answers, the system suggests possible actions. Although many nurses welcome the software, describing it as a safety net for patients and themselves, others find it constraining because it limits their ability to respond or to exercise their clinical judgment (see, for example, O'Cathain et al., 2004). If a tele-nurse chooses to use her own knowledge to provide unscripted advice or extra 'care', she may be stepping beyond what is expected (or required) in her work role. Members of the public generally regard telephone nurse advisory services favourably, but callers may become frustrated if they receive scripted advice that they have already considered on their own, such as consulting their doctor or going to the emergency department (Harris, Wathen, and Fear, 2006). Instead, callers may be looking for confirmation of symptoms, reassurance, or treatment advice, much of which is support that nurses have been trained to give, but may be prevented from providing because of the constraints of tele-nursing procedures (see, for example, Wilson and Hubert, 2002).

The challenge of responding to clients' health information needs within the boundaries of paid employment is also evident in librarianship, the primary purpose of which is to facilitate the transfer of information. Even in this occupation the interpersonal work that underpins successful information exchange is often undervalued or unrecognized. Although librarians can do a great deal to provide lay health information seekers with relevant and useful support, such assistance may not be forthcoming if librarians or their employers do not regard this work as part of the job. For instance, in a study reported by Henwood et al. (2008), staff members working in a public library were limited by operating policies that curtailed the amount of time they are able to provide in response to individual inquiries. As a result, a number of them felt unable to respond appropriately to patrons whose questions arise from serious health concerns. The emotional demands associated with responding to health information inquiries seems not to have been anticipated (and is not encouraged) in the design of services, not only in public libraries but in other library settings as well. For example, in many hospital libraries librarians are expected primarily to help health care providers locate information relevant to their practice and, to a lesser extent, assist patients to locate credible sources of health information relevant to their conditions. However, it is the latter, albeit minor, role that often creates the greatest challenges, especially when newly

diagnosed patients come to the library in search of information to prepare themselves for the future and provide a sense of hope. In these
situations, the librarian may be looked upon, not only as an information guide and interpreter, but also as a source of significant emotional
support, an expectation that pushes well past the boundary of most
librarians' paid work roles. In an increasingly online world, there is a
clear and growing need to connect help seekers with reliable sources
and assist them to make sense of the information available, especially
within the vast, complex domain of health information. Librarians are
professionally educated to address these issues and have the necessary
skills to facilitate or 'add value' to the information exchange process,
both behind the scenes through the organization of information as well
on the front lines, through sophisticated reference or enquiry services.
Yet, despite evidence that information specialists such as hospital librarians can facilitate better medical outcomes (Marshall, 1992), their work
is generally invisible within the formal health system. This is regrettable
since librarians are one of the few professional groups, other than social
workers, whose info(r)mediary role does not involve telling patrons or
clients what to think. Instead, their role is to facilitate access to a range
of information sources, leaving it up to the patron to decide how they
wish to use the information (see Bella et al., 2008).

   What accounts for librarianship's lack of visibility and why are librarians generally absent from the health care info-scape when they could be
at the forefront among those who help lay people to locate, retrieve, and
cope with information that is relevant to their concerns? One reason
may be that, as members of a female-intensive occupation, librarians are
not perceived to possess a great deal of expertise, particularly not that of
a technical nature (Harris, 1992; Harris and Wilkinson, 2004). As a result,
the complex, technologically-based activities that comprise much of the
work of contemporary librarianship are either not recognized or not
attributed to librarians (see, for example, Nilsen and McKechnie, 2002).
Even within the library profession itself, in-depth, interpersonal 'care'
work is not always highly valued, nor is it something for which those
who are responsible for libraries necessarily want to pay. For instance,
reference or 'enquiry' work is a central function within librarianship
and is the role through which care is most directly manifested in the
profession. Despite considerable evidence that library patrons recognize
and value highly the opportunity to consult with librarians who are
friendly, interested, and skilled in determining their information needs
(Durrance, 1989; Ross and Dewdney, 1994), library management practices often position librarians' work in such a way as to minimize the

length of time they are able to spend with library patrons, thereby keeping down costs, as well as users' expectations.

Gender and technology imperatives in health info-mediation

Embedded in the contemporary rhetoric of Western health care is the idea of the 'empowered' health care consumer who is an active partner in health decision making alongside health care providers. The presumed basis of empowerment is information, much of which is available online. Various interests, both public and private, alert consumers to health information via the media or steer them to it online with strategies such as purchasing domain names made of frequently used terms or using meta-tags to guarantee that particular web sites appear early when Internet searchers use commercial search engines. Consumers are also targeted by health providers and insurers who use information 'push' technologies, such as personalized email or automated phone messaging, to deliver reminders to patients about appointments, medication regimes, or post-surgical self-care procedures. Members of the public are also regularly exposed (or subjected) to health-related messages through print-based and televised public health campaigns that encourage citizens to get their flu shots, drink less alcohol, exercise more, and change their eating habits. These campaigns do have an impact, even if their prescriptions for healthy living are not always followed. For instance, one woman we interviewed described her reaction to newsletters she receives from an osteoporosis society: 'They made me develop a guilt complex about my neglectful habits.... they keep saying, you need to quit smoking, you must exercise, you must do this, you must ... If you want to help yourself, you must correct your own personal habits. It's agonizing you know. There's a guilt.'

Promoting the link between health information and consumer empowerment reflects the desire of governments, health management organizations, and employers to reduce health care costs. It is assumed that if citizens, customers, and employees can be encouraged (or feel guilty enough) to keep themselves fit and well, they will rely less on costly medical interventions and reduce demand on the health care system. The empowered consumer is positioned as a person who takes responsibility for her or his own health, is enabled to make health decisions, and is in control and capable of self-care. In other words, empowerment means taking care of oneself, usually at home. However (re)placing responsibility for health in the domestic sphere (as it was in the pre-Victorian era) means that many of the tasks involved in personal health maintenance and illness management will fall to women. As these health duties go home, the public's need for effective health info-mediation will increase.

The work of health info(r)mediation, which is performed by many different people, in many different settings, is usually (and conveniently) invisible to health system 'payers'. With little desire on the part of health management organizations, hospitals, governments, insurers, and others to compensate for the extra care work involved in effective health information exchange, there exists among policy makers and health care managers considerable enthusiasm for technical solutions to meet the information needs of health care consumers. These solutions, such as automated patient information push-out systems and other e-health initiatives, including web-based health information portals, are intended to obviate or reduce the need for paid info(r)mediators. In a review of studies involving health interventions delivered to clients/patients via the Internet, Griffiths et al. (2006) found that cost reduction was one of the most frequently cited reasons for introducing such services. Even in libraries, self-service technologies are being introduced to reduce costs, although they are promoted to library users on the basis that it is 'empowering' for patrons to check their own books in and out, and to find information on their own, without the intervention of library staff.

Regardless of how well such technologies work to package, deliver, and otherwise make health information accessible and available, the health consumer's desire for care, including emotional support and help to interpret complex, uncertain, and upsetting information, is likely to remain. When the mandate and time to care is missing from the jobs of formal health (or information) providers, that is, if the opportunity for listening and meaningful communication is not provided for, recognized, supported, and compensated in the formal system, the burden of this care will continue to drift to the alternative sector: the unpaid and unrecognized domains of women's work. In 1993, Stoller wrote that 'most of the unpaid people who tend to the needs of others, both in sickness and in health, are women' (152). The justification for this allocation of social responsibility is that, since women tend to be lower wage-earners, they have fewer opportunity costs if they take on the job. More than a decade later, Armstrong and Armstrong (2004: 24) observed that,

> when care moves home it usually means care by women because of the assumptions about who should care, the failure to provide

alternative public care, and men's higher wages, which means it makes sense for the women in the house to sacrifice their paid jobs or adjust them to the care work.

The expectation that women should take on unpaid health labour continues to be woven into public policy where, as primary health care systems increasingly emphasize what is euphemistically referred to as 'community participation', 'women's role, although central to virtually all components of [primary health care] is ignored' (Vlassoff and Moreno, 2002: 1714). The assumption that women will assume responsibility for others in their family networks is so embedded within the health care system that several of my women friends have been chided by health service providers for neglecting elderly relatives who live hundreds or thousands of miles away, while male relatives who live closer by face little such pressure.

Just as terms such as 'community-based care' and 'self care' disguise the obligation for women to do more, so too does 'empowerment', since it is usually up to women to locate, interpret, and use the information from which empowerment is to be achieved to support themselves and assist others. For women who have access to the Internet, the technology provides an obvious means of supporting this info(r)mediary role, but it adds to their workload. As Nettleton et al. (2004: 550) point out, 'the information gained is contributing to the lay health work that is routinely carried out by people caring for themselves and others'. Because of the immediacy and continuity of its presence and the vast range of information to which it provides a link, the Internet, along with pervasive social messaging about personal responsibility for health, creates imperatives that are hard for women to resist and add to the burden of their care work.

What are these imperatives? First, the general message that citizens should be responsible for their own health that is promoted, for example, by public health arms of government (see, for example, Bercovitz's (1998) critique of Canada's 'Active Living' policy) and intra-organizational directives from human resources departments, has been recognized and incorporated into common public discourse and into personal identities. As Ziebland (2004: 1783) explains, 'one of the consequences of easier access to health information may be the emergence of a felt imperative to be (or present oneself) as an expert or critical patient, able to question advice and locate effective treatments for oneself'. For instance, several women we interviewed told us that they felt obligated to 'do the research' to support their health needs (Harris and Wathen,

2007). Their comments support Kivits' (2004) claim that the obligation to stay healthy drives 'the endless pursuit of information' (512).

A second imperative extends women's personal responsibility for health to the wellbeing of others, particularly children, partners, and older relatives. As a result, many women feel they must be vigilant and keep up with new information related to illnesses, treatment, health conditions, and strategies for healthy living that are, or might be, relevant to themselves or members of their families (Meadows et al., 2001). We found in our interviews, for example, that women who are not Internet users but who have internalized a sense of obligation about staying informed, 'just in case,' may feel anxious or guilty if they don't wish or know how to search online. They feel pressure to 'learn the Internet' or they turn to others whom they regard to be more Internet-savvy (their children, friends, or sometimes librarians) to search online for them so they can fill their health information obligations, often related to the care of spouses. One woman said, 'I don't want anything to do with this Internet business', yet she asked her daughter to search the Web for cooking tips to manage her husband's high cholesterol. Similar examples appear in Nettleton et al.'s (2004) description of 'reluctant' users of the Internet.

A third imperative comes into play when health gatekeepers learn of or retrieve information that might be relevant to someone in their care network. Often these situations not only require gatekeepers to pass along the information, but also to take some action as a result. Madden and Fox (2006) report that 'e-caregivers' who rely on the Internet during the health crisis of a loved one use it not only to seek advice and support from others, but also to locate services and compare options. The burden involved in information translation and action may be a heavy one. It is not always easy for lay health searchers to make sense of the material they retrieve and digest it into something meaningful and actionable, especially when there is not a lot of clinical certainty or a clear medical solution to a problem. This is evident, for example, in the challenges faced by families who are looking for treatment for autistic children or trying to help parents suffering from dementia.

The way forward

As the volume of health information expands and its complexities increase, the 'need to screen individual pieces for relevance increases, but [the] ability to screen decreases' (Babrow et al., 1998: 17), resulting in 'continuous insecurity and uncertainty' (Kivits, 2004: 524). In the face of such uncertainty, it is not surprising that when individuals take on more responsibility for their own health, the process of interpreting and using health information becomes more difficult and intense. For all the Internet can do to facilitate access to useful health information and connections with others who are able to offer support, the pressures to stay health-informed and to apply health information are adding to the stress women already experience with respect to family health. Facilitating access to online health information is valuable, yet for many women it further reinforces traditional gendered roles.

In considering the future, it is important to bear in mind that information, no matter how well packaged or conveniently delivered, is not a proxy for care. When access to care is limited and professional care-givers, particularly physicians, are unable to spend the time necessary to ensure that patients and their advocates not only comprehend but are able to cope with the implications of health information, the development and support of alternative methods for providing health info(r)mediation should be on the policy agenda. One option is to encourage the development of strong connections between health services and public libraries and to provide the libraries with sufficient resources to enable the development of up-to-date health collections and to employ librarians who have the appropriate training and experience to work with lay health information seekers. Another option is to develop the role of information navigators such as patient-centreed librarians in health care settings. To be effective, these roles must not only be recognized by health care practitioners, but those who perform the work must be adequately compensated and included in the design of patient support services. Relying on a combination of web sites and women's unpaid labour is an unrealistic remedy to a growing problem. If we are to sustain a public health imperative that requires 'good' citizens to stay healthy and, in order to do so, keep themselves informed, we need a concomitant imperative to recognize and invest appropriately in the work of the health info(r)mediaries to whom they turn for support.

5
Nursing Technologies? Gender, Care, and Skill in the Use of Patient Care Information Systems

Zena Sharman

A systematic review of the impact of health information and communication technologies (ICTs) on medical care notes that ICTs 'do not, in and of themselves, alter states of disease or of health' (Chaudhry et al., 2006: E-18). The authors emphasize the importance of how ICTs are used and the context in which they are implemented in their discussion of the potential impacts of such technologies. This book is similarly concerned with understanding ICT use in context, though it adds an additional dimension—that of gender—to our exploration of the subject. This chapter examines how gender mediates the way nurses use and think about ICTs in the context of their work as caregivers. Nursing work has traditionally involved a great deal of highly skilled carework, yet it increasingly requires the use of ICTs, some of which have a primarily administrative function. This shift, which is related to structural changes in health care that have transformed the nature of nursing work, has created a unique set of gendered tensions as a feminized profession grapples with its relationship to masculinized technology and feminized administrative work. This chapter explores these tensions through reference to empirical data from a qualitative case study of Emergency Department nurses' interactions with and perceptions of a computerized Patient Care Information System (PCIS), an administrative technology designed for the purpose of collecting standardized patient data. It discusses how nurses working in a context of health system restructuring define their skilled caring work in contrast to both low-status feminized administrative work and depersonalized masculine technology. In this way, the PCIS becomes a site of gendered tension and resistance as nurses articulate the role of technology in nursing work practice.

ICTs in context: Health care restructuring in Canada

We must examine the wider policy environment in order to understand ICT use and how processes of health care restructuring shape nurses' work practice. Restructuring of the Canadian health care system is rooted in the transition from an interventionist welfare state to a neoliberal paradigm that began in the early 1970s. The neoliberal paradigm emphasizes market mechanisms such as competition and private delivery of services (Armstrong and Armstrong, 2001b). Under this paradigm, health care is conceptualized as a business and patients/citizens as consumers. Although the Canadian health care system remains universal and publicly administered, it increasingly conforms to the 'care-as-another-business model' (Armstrong and Armstrong, 2002). This model emphasizes managerial control, flexible work assignments, and delegation of tasks to lowest-cost care providers. It reflects a pattern of structural changes that have occurred over the last two decades throughout health care systems in North America and Western Europe (Malone, 2003), which include the shift of caregiving from institutions to communities, shortened hospital stays, increasing casualization of health care work, use of private sector management techniques (e.g., total quality management), and an emphasis on standardization (of clinical practice, patient classification, and data). Health system restructuring has had significant implications for the everyday/everynight work of nurses and other health care providers.[1]

The care-as-another-business model shaped responses to the fiscal crisis that occurred in the Canadian health care system in the early 1990s when an economic recession was followed by $20 billion in cuts to federal transfer payments (Ostry, 2006: 204). Canadian provinces used transfer payments to fund social programmes such as health care, so the funding cuts had a drastic impact throughout the country's health care system. Canadian hospitals were not immune to the effects of these cuts. Between 1989 and 2003, there was a 40.4 per cent decrease in the number of hospitals and a 36.1 per cent decrease in the number of hospital beds in Canada (Ostry, 2006: 208). This loss of acute care capacity changed health care delivery both within and outside of hospitals. The smaller number of hospital beds meant that they became a resource reserved only for the sickest patients. The reduced availability of hospital beds coupled with advances in medical technology led to the shortening of inpatient stays and increased use of day surgery. These changes intensified the work of hospital-based health care providers who now had to care for very sick patients under strict time pressures. These changes also intensified the work of formal and informal caregivers in the community as early discharge and the limited availability of hospital beds increased the number of people requiring home care. Paid and unpaid health care providers are predominantly female. The intensification of their work as a result of restructuring reflects Neysmith's (2000) observation that restructuring has a disproportionate effect on women because they provide a disproportionate amount of care. Some authors argue that restructuring has changed nurses' work so significantly that it is prompting nurses to leave the profession and is thus causally related to the North American nursing shortage (White, 2003).

Increased use of computerized patient care information systems is another outcome of health system restructuring. These systems perform two functions in health care work: they are used to *accumulate* data gathered throughout a patient's trajectory and they *coordinate* events in disparate locations, which are undertaken as part of efforts to support diagnosis and continuity of patient care (Berg, 2004). PCISs were first developed in the 1970s, but did not achieve widespread use until the 1990s. Balka (2003a) notes that the Canadian health 'info-structure' developed as part of the restructuring process, which reflects a broader focus on improving the efficiency and effectiveness of the health care system through implementation of practices such as evidence-based decision making. Policy-makers and hospital administrators have looked to PCIS as a means to cut costs and increase efficiency through improved resource management, data collection (for measurement of outcomes such as waiting times), and accountability (Choiniere, 1993). The growth of PCIS is also linked to the emergence of evidence-based practice as the dominant paradigm in health care. These technologies are thought to facilitate the use of standardized charting practices, patient classification systems, and clinical protocols, and support error reduction and measurement of patient safety (Berg, 2004).

Decision-makers implement PCIS in order to improve the efficiency and effectiveness of the health care system. There has been widespread uptake of these technologies, a very costly endeavour that has changed work practices throughout the health care sector. However, the evidence suggests that the proliferation of these technologies is related to what they represent rather than to a strong track record of success. It is estimated that somewhere in the range of 30–75 per cent of health information systems fail (Pratt et al., 2004).[2] These failures sometimes result in loss of life or enormous losses of public monies (Beynon-Davies, 1999; Craig and Brooks, 2006). Patient care information systems often fail because the model of health care work inscribed in them does not match the reality of health care work practices (Berg, 2004). The proliferation of these systems in spite of the lack of evidence for their success is puzzling when considered in the context of the broader preoccupation with evidence-based decision-making. Wagner (1993) explained this phenomenon in terms of what these technologies signify rather than what they actually do. She suggested that computers 'act as powerful images of efficiency' (298). No longer a means to an end, efficiency has become an end in itself in processes of health care restructuring (Stein, 2001). The preoccupation with efficiency goes hand-in-hand with efforts to cut costs, standardize practices, and increase accountability (largely through data collection and reporting). Computerization and restructuring of the health care sector are linked processes with implications for health care work practice.

Gender, carework, and ICTs

Health system restructuring and related processes of PCIS acquisition in health care have transformed nursing work. For example, nurses are increasingly responsible for administrative work and supervision of lower cost care providers, which detracts from time for hands-on patient care (Malone, 2003). Staff shortages mean nurses sometimes have to perform tasks normally done by ancillary workers, such as feeding patients or cleaning hospital rooms (Armstrong et al., 2006). Nurses also spend more time doing data entry due to increased use of PCIS (Choiniere, 1993). These shifts have fundamentally changed the nature of nursing work.

Nurses seeking to articulate their professional identity in this context must contend with two gendered social constructions: the feminization of carework and the masculinization of technology. Nurses also have had to contend with gendered definitions of skill and the related devaluation of their work as caregivers. This has implications for how nurses define their expertise and their professional identities. Throughout their professional history, nurses have aligned themselves with and against caring, as well as aligned themselves with and against technology. Nurses have also sought to situate themselves within the health care system's occupational hierarchy by differentiating their work and expertise from that of higher status physicians and lower status ancillary workers.

Nursing and the feminization of carework

Caring has a dual nature in that it demands both labour and love (Graham, 1983). The dual nature of caring is reflected in the distinction between *caring for*, which 'refers to the instrumental and tangible tasks involved in caring', and *caring about*, which 'encompasses its expressive and affective dimensions' (Baines et al., 1991: 15). I utilize the term care-work in order to reflect the multiple dimensions of caring (Zimmerman et al., 2006: 3–4). How individuals experience the multidimensionality of carework is mediated by social location, which describes how individuals and groups (e.g., women workers) are 'affected differently by social relations of inequality such as gender, race, ethnicity, immigrant status, disability, class, and age, as well as their intersections' (Vosko, 2006: 459). A wealthy, white man in good physical health will experience the socially constructed obligation to care differently than a poor woman of colour suffering from a chronic musculoskeletal disorder. There are myriad experiences of carework, each one mediated by social location.

One dimension of social location that influences carework is gender. Graham (1983: 21) observed that 'what counts as caring is determined as much as by who does it as by what is done'. In Western culture, the social construction of femininity reinforces the association between femaleness and caring. Women are socialized to care for others and caring is perceived to be an inherently feminine trait (Cancian and Oliker, 2000). As caring is thought to come 'naturally' to women, it is perceived to require little skill, which reflects how gendered assumptions shape the social construction of skill (Phillips and Taylor, 1980). As a consequence of such assumptions, carework is undervalued and rendered functionally invisible by its taken-for-granted status (Star and Strauss, 1999). It is a form of invisible work that is only apparent when it is not done (Graham, 1983).

The feminization of carework has consequences for the gendered division of paid and unpaid labour. Women are primarily responsible for the care of partners, children, and older people. This is reflected in Canadian women's disproportionate share of unpaid work in the home (4.3 hours per day), almost twice the amount of time spent by men (2.5 hours per day) (Statistics Canada, 2006: n.p.). It is also reflected in the gendering of occupations. Women are overrepresented in the 'caring professions', such as social work, teaching, and nursing, in which women 'become involved in housekeeping tasks on behalf of society at large' (Graham, 1983: 16). Feminist nursing historians describe the history of their profession in terms of a transition from unpaid care-work in the home to paid carework in private duty and in hospitals (McPherson, 1996). In 2004, 95 per cent of registered nurses in Canada were women (Canadian Institute for Health Information, 2006: 207). Nursing is a highly feminized occupation that exemplifies women's socially constructed obligation to care (Reverby, 1987).

The association with feminized carework has implications for the visibility and value of nurses' work, as well as perceptions of their skills. This is of particular relevance given the changes to nurses' work that have occurred as a consequence of health system restructuring. Nurses have utilized the association between nursing and caring in different ways throughout nursing's professional project, a process Witz (1992: 64) defines as 'strategies of occupational closure which seek to establish a monopoly over the provision of skills and competencies in a market for services'. Nurses have sought to differentiate themselves from physicians, who dominate the medical hierarchy, and from ancillary workers, such as care aides or clerical workers, who are nurses' subordinates (Witz, 1992). At times, nurses have aligned their profession with feminine caring as a means to distinguish their expertise and unique contribution to health care (Sandelowski, 2000). Nurses have also de-emphasized caring in order to highlight their technological aptitude, specialized knowledge, and efficiency (Sandelowski, 2000). Such strategies have been controversial because they sometimes blur the boundaries between nursing care and the low-status administrative tasks typically performed by clerical workers (Bowker and Star, 1999). These varied strategies demonstrate that there are multiple ways to frame the gendered associations between nursing and carework.

Nursing and the masculinization of technology

Gender has also shaped the ways nurses have utilized technology in service of their professional project. Wajcman (2006: 780) observes that gender is 'constitutive for what is recognized as technology, and gendered identities and discourses are produced simultaneously with technologies'. In Western society, technology is masculinized and technical competence is assumed to be an inherently masculine trait (Grint and Gill, 1995). Technological expertise is highly valued and perceived to be highly skilled, though, like carework, these appraisals depend on both user and context of use. A male and female nurse using the same heart monitor on the same patient may be differently perceived according to gendered assumptions about his or her technical abilities.

Nursing and technology are inexorably linked, for nurses have always utilized a range of technologies (e.g., thermometers, cardiac monitors, respirators, hospital beds) in their work (Sandelowski, 2000: 1). As such, it is not surprising that nurses have at times highlighted the technical aspects of their carework in an effort to demarcate their areas of expertise. Nurses have turned to technology to mitigate their association with unskilled 'bodywork', the intensely physical aspect of carework (e.g., bathing, toileting) (Sandelowski, 2000). Technology has functioned as a means to validate nursing knowledge, as in the case of electronic foetal monitoring, which makes 'visible the accuracy of the nurse's "intuition" ' (Sandelowski, 2000: 154). Other nurses have emphasized technology's association with science, progress, and professionalism in order to strengthen the scientific basis of nursing (Wagner, 1993). This perspective is reflected in efforts to standardize nursing practice and in the development of specialized fields of expertise such as nursing informatics (Ball et al., 2000).

Nurses have also opted to turn away from technology. Such strategies often depict nursing/touch and technology as oppositional paradigms of care, a duality that reflects gendered assumptions about carework and technology (Sandelowski, 2000: 9). This duality is evident in Malone's (2003) distinction between 'distal' nursing (rationalized, standardized, and high-tech) and 'proximal' nursing (individualized, context-bound, and 'high-touch'). Technology becomes a tool for the subordination of nursing through erasure of nurses' skills and expertise (Sandelowski, 2000). This erasure is often achieved through 'textualization' of nursing knowledge and practice, in which 'textually mediated forms of organising health care render what were once professional judgments made by individual nurses into organisational judgments made through complex documentary (including electronic data) processes' (Campbell, 1992: 752). A reorientation from a standpoint centred on care to one centred on efficiency accompanies this process.

Nurses' resistance to administrative technology

Resistance to the textualization of nursing knowledge and practice is an implicit theme in studies of the implementation of administrative technologies (e.g., patient care information systems, care planning software, automated drug-dispensing systems) in nursing. The studies discussed here all describe technologies implemented as part of processes of health system restructuring. The evidence indicates that nurses' resistance often takes the form of 'resistive compliance' (i.e., attempts to minimize or 'put off' use of the systems and extensive criticism of the systems) (Timmons, 2003b). Nurses criticize the systems for being time-consuming to use and argue that the requirement to use administrative technologies in their work reduces the amount of time they are able to spend with patients (Choiniere, 1993). Nurses also perceive these systems as unreliable, as they often crash or are taken offline for maintenance (Timmons, 2003b).

Timmons (2003b: 257) notes that nurses' resistance to such technologies is 'as much about the ideas and ways of working that the systems [embody] as it [is] about the actual technology being used'. It is apparent that these technologies do more than change the nature of nurses' work—they challenge the very core of nurses' skilled carework. This is reflected in nurses' concerns that over-reliance on technology might diminish their assessment, communication, and other clinical skills, or increase the risk of error (Choiniere, 1993). In the face of such concerns nurses assert the 'value of their experienced human judgement in protecting patient safety' (Novek, 2002: 400). Nurses also seek to distinguish between what they describe as skilled nursing care (*patient care*) and unskilled administrative work (*paper work*) (Choiniere, 1993). In this context computers come to signify 'the lower status feminine task of typing', while also being simultaneously perceived 'in masculine terms as "taking over" or controlling' (Henwood and Hart, 2003: 257). Administrative technology becomes a point of intersection for various aspects of nursing's professional project. It detracts from patient care by placing a growing burden of clerical work on nurses, thereby cementing its association with tasks normally performed by subordinate administrative workers. At the same time, technology signifies male dominance and the erasure of nurses' clinical skills. The evidence suggests that nurses reaffirm the value of their carework in the face of such threats to their professional identity.

Case study: Gender, ICTs, and skilled carework in the emergency department

The following case study is used to illuminate the way gender mediates nurses' perceptions of ICTs—specifically, administrative technologies such as the PCIS examined here—and carework. It illustrates how the gendering of tasks and tools shapes nurses' professional identities as they strive to situate themselves within the hospital's occupational hierarchy. Material here is based on a study that examined how nurses perceive and use ICTs in the context of their carework.[3] It looked at a range of technologies, but was primarily focused on a PCIS. The research site was the Emergency Department at Vancouver General Hospital (VGH), a large urban tertiary care centre in Vancouver, British Columbia, Canada. Hospital administrators identified implementation of a PCIS as a top priority in 1991 (Balka, 2003a). After the hospital was unable to obtain provincial funding for the PCIS, VGH paid for the system by entering into a controversial public–private partnership with a telephone company. The PCIS was implemented in 1997 (with the promotional slogan 'Data Sharing Leads to Patient Caring'). It was the target of criticism from health care workers' unions, who contended that users had not been adequately involved in system development and that the PCIS was disruptive to existing work processes (Balka, 2003a).

The PCIS came into being during a period of extensive health care restructuring in British Columbia. The province lost a significant amount of its acute care capacity between 1991 and 2001 (Cohen et al., 2005). British Columbia underwent three successive phases of health system regionalization during the same period. This process involved a massive reorganization of health system governance at the local, regional, and provincial levels. Balka (2003a) notes that the same period was also characterized by significant interest (and investment) in health information technology at the provincial level. These events formed the context for the implementation of the PCIS at VGH.

Data collection consisted of observations of staff and interviews with nurses, and focussed on how individual nurses and clerks and groups of workers utilized the PCIS and other technologies, as well as general work processes in the Emergency Department. Interviews elicited nurses' perceptions of the PCIS, their definitions of care, and their perceptions of the relationship between caring and technology. Data were primarily collected in the triage area, which is a central location in the Emergency Department staffed by highly trained nurses whose primary task is the assessment of incoming patients. Triage nurses decide how quickly and by whom patients need to be seen. They are responsible for entering a limited amount of patient information into the PCIS for every patient they see and often use the system for information retrieval (e.g., to look up a patient on another hospital ward). A small number of admitting clerks work at a row of desks adjacent to the triage area. They are responsible for registering patients' demographic information into the PCIS. Although these two groups of workers were physically proximal, there is not a great deal of professional interaction between them.

The case study with the Emergency Department nurses reveals the multiple ways that gender shapes their work practice, perceptions of technology, and professional identities. The PCIS is integrated into nurses' work practice to the extent that they use it to accumulate and coordinate information about patients. However, nurses articulate a clear boundary between administrative technologies such as the PCIS (and related tasks) and their core work as skilled health caregivers. Nurses also emphasize the importance of their embodied caring skills and relationships with patients, which they contrast with cold, mechanical technology.

Work practice and the PCIS

The PCIS is integrated into nurses' work practice in that it performs the accumulating and coordinating functions described by Berg (2004). Triage nurses play a central role in data accumulation in that they are responsible for patients' initial registration into the system. The nurse on duty conducts a brief physical examination of each new patient. Her findings form the basis for a PCIS entry noting the patient's name, complaint, and time of arrival. Nurses also write down patient information in paper-based patient charts, as the hospital used both paper and electronic records during the study period. Admitting clerks subsequently add additional demographic information (e.g., address, occupation, next-of-kin) to the patient's electronic record. The contents of a patient's PCIS record will change according to an individual's status and physical location inside or outside the hospital (which might be listed as waiting to see a physician, admitted to the Emergency Department, moved to a specialized ward, discharged, or deceased). The system makes it possible to transmit such patient information over time and across organizational boundaries.

Emergency Department nurses describe this function in terms of 'keeping track' of patients, as they use the PCIS to follow a patient's trajectory through the Emergency Department and the larger space of the hospital, as well as to follow up on the results of procedures or laboratory tests a patient may have undergone. Keeping track of patients involves situating them in the space of the PCIS. Nurses frequently refer to patients as being inside or outside the system, as signified by comments like, 'He's in the system now' (Nurse D., April 2003) or 'You're out of our system' (Nurse X., February 2003). Patients exist not only in person but also in the system; they are fixed in a 'machine space' that bounds and structures activities (Tellioglu and Wagner, 2001).

Nurses also use the PCIS to coordinate activities. The system displays data and represents the Emergency Department in a way that makes it possible to manage people in space (by assigning and/or shifting beds, prioritizing patient care based on acuity levels, and so on). In addition to the registration screen and lab results display, one of the most frequently viewed screens is the Emergency Department census. The Emergency Department census displays an unalphabetized list of names and selected patient information for all patients in the department. Although it is a text-based interface, triage nurses use the census as a means to visualize the space of the Emergency Department and the people in it, and to communicate with colleagues and co-workers throughout the hospital. This display screen figures into nurses' conversations about patients as they strategize about patient care. Triage nurses frequently consult the census to gauge current conditions in the department and use it as a 'prop' in discussions about patients and the allocation of beds. When talking about a specific patient, nurses will sometimes highlight the person's name on the screen. The PCIS thus functions as both a source of accumulated patient data and a tool for the coordination of activities in the Emergency Department, what scholars working within a computer-supported cooperative work perspective might refer to as an 'ordering system' (Schmidt and Wagner, 2004).

Skill, administrative work, and nurses' professional identity

Although the PCIS is integrated into the triage nurses' work practice, use of the system is not reflected in how nurses conceptualize their professional identities and responsibilities as skilled caregivers. The system is marked as separate from or not implicated in patient outcomes: 'I don't give it [PCIS] a great deal of thought...because it's not directly related to the outcome of patients' (Nurse P., April 2003). Nurses do acknowledge that certain technologies (e.g., a heart monitor) contribute to patient care, but these are largely discussed in terms of clinical care. As an administrative technology, PCIS stands apart from the clinical technologies nurses routinely use in their carework. The particular positioning of administrative technologies can be attributed in part to the controversial status of administrative work in nursing.

Emergency Department nurses assigned low priority to administrative tasks. This is a byproduct of their heavy clinical workload, but it also reflects attitudes towards the place of administrative work in nursing practice. Although they complete the requisite administrative tasks, nurses highlight patient care, not *paper work*, in their work processes and descriptions of their work. The nurses' first priority is always patient care, so administrative tasks sometimes fall to the wayside. When possible, nurses try to avoid using the PCIS. This strategy sometimes results in the downloading of administrative responsibilities to lower status clerical workers:

> I try to use [the PCIS] as minimum as possible...so I do what I need to do for triage, for admitting. We're meant to discharge patients out of the system, but I don't because it's a reasonably lengthy process. It's not long, but in terms of how much time we have, it's long enough, so I tend to sort of rely on the unit clerks to keep it updated.
>
> (Nurse P., April 2003)

Nurses are supposed to complete paperwork in order to discharge patients, but often choose, or are compelled by their workload, to leave the task for one of the Emergency Department's unit clerks to complete. Nurses also emphasize the distinction between their work and the administrative work performed by clerks. As one nurse told a patient, 'I'm here for your health. They're [the admitting clerks] here for all that computer stuff' (Nurse A., February 2003). Over the course of my observations in the Emergency Department, I heard nurses and clerks utter many variations on this statement. It reflects nurses' efforts to distinguish their carework from low-status, 'unskilled' administrative work. It is part of a strategy of occupational closure that involves establishing a monopoly over nursing care by emphasizing distance between their carework and the clerical tasks normally performed by ancillary workers.

Administrative tasks connote feminized clerical work. This may explain why nurses, already employed in a feminized profession, seek to distance themselves from their clerical counterparts. It is a distinction partly premised on gendered definitions of skill. The ability to effectively use and navigate a system like PCIS requires users to draw on a repository of tacit knowledge as they engage in a significant amount of invisible work. However, the masculinization of technical know-how renders women's work with technology unskilled, even when they utilize sophisticated technologies to perform complex tasks, because skill is gendered. The assumption that women are 'naturally' adept at caring leads to nurses' feminized caring work being similarly stripped of skill. Nurses do not subscribe to this image of their work, however, for they are best positioned to acknowledge the skill caregiving requires. By distancing their professional identity from administrative responsibilities, nurses assert a desire for professional respect and recognition founded on a revaluation of skilled caring work.

The importance of embodied caring

In addition to distinguishing their carework from administrative tasks, triage nurses emphasize the uniqueness of their embodied caring skills in the context of a highly technologized environment. This perspective is evident in how nurses conceptualize technology, which they tend to define in contrast to caring. Technology is 'cold and mechanical' and 'you don't correspond it with human caring' (Nurse C., May 2003). Technology is about 'productivity', and 'I don't think productivity is about caring' (Nurse P., April 2003). Technology is scary because of 'the lack of human contact ... It feels very depersonalized or dehumanized' (Nurse M., March 2003). This stands in sharp contrast to caring, which involves 'listening, talking to somebody, letting them realize that they're being listened to, trying to understand their needs and anticipate them...just being there for somebody' (Nurse W., March 2003).

The nurses I interviewed emphasized emotional connection in their definitions of care. As one nurse explained, 'the most meaningful thing is the personal relationship that I have established with my patient, their family and my fellow nurses. It's not how well I use my technology' (Nurse M., March 2003). A very important part of caring is 'just to be able to hold a patient's hand while they're gasping their last breath...you can do a lot of things, but I think that the human contact, the personal touch, that is the most important part' (Nurse W., March 2003). The same nurse explained,

> I don't think anything's going to take over the human touch. I mean, I don't think we can design, well, maybe two hundred years from now we can design a computer that does everything but can it really sit there and hold your hand when you're sick?
>
> (Nurse W., March 2003)

The emphasis on touch and emotional connections underscores the performance of emotional labour in nursing. 'True caring...is to really be emotionally involved. And that's hard, that's hard on me' (Nurse M., March 2003). Emotional labour is hard work. Like carework, it is feminized and undervalued.

Nurses also emphasized the importance of their embodied skills and caring abilities. Technology is a 'good tool', but 'it cannot be the only tool and it can't be the strongest tool' (Nurse M., March 2003). The strongest tools are a nurse's own embodied skills and her caring ability. As one nurse explained,

> Baseline is you always rely on yourself. Machines aren't perfect. You know, a flat line on a machine doesn't necessarily mean your patient's flatlined. You've always got to look at the patient, but technology definitely makes it a lot easier. There's definitely a place for that.
>
> (Nurse K., April 2003)

Nurses emphasize the importance of their own assessment skills because monitoring and the use of clinical technology have 'to be coupled with our ability to still physically assess somebody and still be able to interact with that person' (Nurse M., March 2003). If nurses 'keep abusing technology...we'll miss the basic assessment skills that we have. We rely too much on the machine' (Nurse J., March 2003).

Nurses trust their embodied caring skills more than clinical technology because, unlike technology, their caring skills do not have the same built-in capacity to malfunction. A caregiver may become exhausted or find it difficult to care for a particularly challenging individual, but the capacity to expertly understand the complexities of her caring ability lies within her. Clinical technology is much more opaque, and nurses rely on their embodied caring skills to validate or challenge what it tells them: 'You can look on a monitor and say, "Gee, that rhythm doesn't look right," and you haven't put your hand on the patient's pulse. You may realize that it's the monitor that's the problem, not the patient' (Nurse M., March 2003). I observed this phenomenon on a number of occasions, when, for example, an electronic blood pressure monitor gave a reading that the triage nurses did not agree with. In such cases, the nurses used a manual blood pressure machine to get a second reading, which they thought was more accurate than that given by the electronic machine. Technology is not fool-proof and it is not always reliable. Although they routinely use clinical technology in their work, nurses rely first and foremost on their embodied caring skills and their ability to use all five senses when physically assessing a patient.

Conclusion

The Emergency Department described in this case study is a nexus of gendered social relations. Health care organizations are highly sex-segregated, as Balka illustrates in her chapter on gender, information technology, and health sector work (see Chapter 5). This case study represents a valuable opportunity to examine women's work (in this case, work that is both feminized and done primarily by women) in a contemporary health care setting. The gender dynamics between groups of women workers in this context are particularly compelling, in that nurses actively created hierarchical gendered identities demarcated by responsibility for and association with administrative work. In doing so, they distanced themselves from their lower status clerical counterparts. Is this a missed opportunity for solidarity among gendered users of technology? The Emergency Department is also a site in which wider patterns of political and economic change are manifested. It is in health care settings such as this one where the effects of restructuring move from the realm of the theoretical to the realm of the real, and where the challenges and opportunities of situated ICT use arise.

This becomes apparent when examining the experiences of the nurses negotiating their professional identities in the context of changing work practice.

By examining ICT use in context, we are able to gain a much deeper understanding of the complex and multifaceted ways that gender mediates work practice and professional identities in health care. This is exemplified by the experiences of nurses in the case study, who use the PCIS to accumulate data and coordinate activities yet do not include the PCIS in their descriptions of the important elements of patient care. This is because the system for them represents both low-status feminized administrative work and depersonalized masculine technology, which they seek to distance themselves from. In response to the encroachment of PCIS technology on their carework, nurses working in a context of health system restructuring utilize a dual strategy of occupational closure in articulating their professional boundaries. They define their work in contrast to administrative work. Although their job design is such that they cannot entirely avoid administrative tasks (e.g., data entry in the PCIS), nurses actively mark out these tasks as separate from their central purpose of caring for patients. Nurses also define their work in contrast to depersonalized masculine technology, which is represented as antithetical to caring. Instead, nurses emphasize the importance of their embodied assessment skills and emotional connections with patients. My findings thus reflect Timmons' (2003b) observation that resistance to technologies like the PCIS is not just about the technologies themselves. It is also about the ideas and ways of working embodied in such technologies.

These findings have implications for the future of the nursing profession. If current patterns of health care reform continue, nursing may become increasingly technologized and standardized. Hands-on carework may be downloaded to lower cost care providers such as nurses' aides, leaving nurses with a greater administrative or managerial workload and less time for direct patient care. Efforts to implement evidence-based practice might create further distance between nurses and their core caring values, since these practices are intangible and difficult to quantify. How might these changes shape nursing work practice and nurses' professional identity? The answer to this question might vary by generation of nurses. Younger, more technologically savvy nurses trained in the principles of evidence-based practice may adapt more easily to this environment than their senior colleagues. Nurses might resist these changes, calling for a revaluation of carework as a form of skilled professional practice. Whatever the answer might be, the nursing profession must continually negotiate gendered fault lines as it determines its position at the nexus of carework, technology, and administrative tasks.

Notes

1. Smith (2002) uses the expression everyday/everynight worlds as a reminder that women's work does not merely take place during the day.
2. Information system failure is a multi-faceted concept which encompasses systems that do not work, perform poorly, are hardly used, fail to meet objectives or expectations, or go over budget (Beynon-Davies, 1999).
3. The Social Sciences and Humanities Research Council of Canada provided financial support for this research through funding for a project titled From Work Practice to Public Policy: Case Studies of the Canadian Health Information Highway (Dr Ellen Balka, Ph.D., principal investigator, grant #410-2000-1096).

6

Gender, Information Technology, and Making Health Work: Unpacking Complex Relations at Work

Ellen Balka

Virtually all health sector work requires a high degree of interaction with technologies, ranging from medical devices such as defibulators to electronic medical records. Although this chapter focuses on information technologies specifically, the definition is a broad one and includes any technologies that provide a means through which health information is made visible to, and circulated between, staff and practitioners who then act on that information as they manage patients' health. Within this definition, heart monitors, which convey information about heart rate and heart rhythms over time, alerting practitioners to problems requiring action may be thought of as information technologies. More conventional definitions of information technologies would include electronic patient record systems which record test values and data about patients over time, in support of diagnosis and care planning. When successfully embedded in work practices, health information technologies can support managerial, clinical, and/or administrative staff as they work together with patients to produce health.

Studying the use of health information technologies by health sector workers brings numerous intersecting issues to light. Such technologies may play a role in altering work organization, professional practice, and the movement and use of information among members of care teams who cooperatively manage patients' health. Where technologies are introduced without due care being given to the work practices involved in health care, there can be resistance to, and/or sub-optimal use of the technologies. In this chapter, I explore information technology implementations that presented problems in relation to work practices and analyse the problems from a gender perspective. Most health sector workplaces are characterized by a gender-segregated labour force, where certain occupations (such as nursing) are filled predominantly by women, while other occupations (ranging from medical specialties to maintenance) are filled largely by men. Results from fieldwork undertaken through the Assessment of Technology in Context Design Lab[1] since 1997 suggest that gender dynamics can prevent technology from being implemented to support the delivery of adequate patient care. In this chapter, I illustrate how information technologies contribute to the gendering of work in the health sector, how gender relations in health sector work come to bear on how technologies are implemented in health sector workplaces, and how an understanding of how gender works in health care contexts can assist in understanding (and ultimately, resolving) problems with technology implementations in the health sector.

Ericksson-Zetterquist (2007) has suggested that gender and technology remain in a complex relationship in the organizational context. In this chapter I explore the complexity of gender–technology relations in health care organizations, by drawing on insights from several areas of scholarship, including science and technology studies (STS), gender studies and organizational studies and analysis. In particular, I show how theoretical insights from STS and organizational studies, when combined with Harding's gender framework (Harding, 1986), can be used to help make sense of the complexities of gender–technology relations in health sector workplaces and how this understanding can, in turn, help us make sense of the challenges experienced in many health information technology implementations. I begin by introducing four vignettes that each describe a problematic IT implementation. I then draw systematically on these three substantive bodies of literature—STS, gender studies, and organizational studies—to analyse these problems in more depth, identifying the contribution that each body of literature can make. I conclude by reflecting upon how such insights might be used to improve the practice of IT implementation in health sector workplaces. The empirical material that I draw on for this analysis was gathered during observationally based field work that was conducted in hospital settings over a four-year period between 2003 and 2007.[2]

IT implementation—four vignettes

Vignette 1: The Cord in the Drawer
After renovation of a hospital unit, a cord which connected an automated drug dispensing machine (a cabinet that could

be accessed only through a keyboard and touch screen) to a refrigerator (the access to which was also controlled through the automated drug dispensing machine) had not yet been fixed permanently into place. It ran between a drawer and its lock, which prohibited the drawer from closing properly. This drawer held narcotics and was supposed to be locked at all times. The aftermath of the renovation had made work for staff more chaotic than normal but, still, narcotics were supposed to be under lock and key. Although many of the staff had to retrieve narcotics from the normally locked drawer and would have observed that it could not be locked, no one told the unit manager or contacted the maintenance department to report the problem.

Vignette 2: The Keyboard Trays
In conjunction with the same hospital renovation identified above, new computer workstations were built to accommodate all the computers in a building. Each was outfitted with an under-the-desk keyboard tray that staff were supposed to be able to tuck away when the keyboard was not required, and retrieve when the keyboards had to be used.

Although the unit clerks and other staff were grateful to have been given these ergonomic keyboard trays, staff members from several units became frustrated because the keyboard trays did not seem to work properly. Staff members had trouble getting the keyboard trays in and out, and the trays led to at least one injury. Having received numerous reports from users about problems with the keyboard trays, our research team[3] conveyed the frustration and complaints to the operations team, who suggested, over several weeks, that the problems were resulting from users' inability to properly operate the keyboard trays. In the face of ongoing complaints, the operations team and the unit that had selected the keyboard trays agreed to let research staff provide on-site training to users. It was agreed that the unit responsible for the trays would train research staff, who in turn would pass information about proper use of the keyboard trays on to end users. In the process of training research staff to support end users, it became clear to the department responsible for the trays that they did not work properly, and further investigation determined that the entire batch of 80 trays was defective. It eventually became clear that no one had checked the order for the trays when they came in and that the installed trays were not only defective, but also were not what had been ordered. The trays were removed and replaced once information about the deficiency had been documented by the unit responsible, and the information had been given to the senior operations team.

Vignette 3: The Wireless Call System
A wireless call system was introduced on a hospital unit. The purpose of the system was to integrate several different types of paging, alert, and calling systems. The new system meant that patients calling the nurses via an over-the-door light system, doctors calling nurses to a particular place or patient, and other types of alerts and alarms (e.g., those indicating that patients had attempted to get out of their beds) could all be routed through to wireless handsets carried by staff around the units as they went about their work. The handsets also made it possible for a nurse to answer a patient's page and speak to the patient from wherever on the unit the nurse happened to be.

At the start of each shift, staff had to log on to the call system and indicate which patient rooms they would be responsible for during the shift. They were to clear their room assignments from their handset at the end of each shift so that the handsets would be ready to be programmed for use by the next shift. On the first day of the implementation of the system, one of the nurses signed out her telephone and turned it on. Immediately it began ringing. The display showed a room number to which the phone had not been assigned. The call was cancelled but throughout the day, the phone continued to receive these phantom calls from a room that was empty. Over the next few days, this happened several times to other staff. The research team listened to complaints, noted the time, phone, and room numbers involved, and double-checked that phones were correctly assigned on the console at the nursing base. The phantom calls were not taken very seriously by the vendor representative who said they would probably disappear again, but eventually he was forced to check the system's software logs. He found an error in the configuration which made the system attempt to call the phones up to one hundred times if calls had been forwarded to a handset while it was turned off (as happened frequently when staff took breaks).

Vignette 4: Electronic Triage

A decision was made by one of the emergency room (ER) doctors to computerize the triage function in an emergency room. Nursing staff were sent on a course to learn to use the program so that they could teach others how to use it when they returned. The program's features included decision support for triage nurses as they assigned a score indicating how urgently a patient needed to be seen. It also prompted nurses to collect specific information (such as vital signs), and provided pull-down menu choices of symptoms patients might present with. Although there were some clear advantages to using the electronic triage system (such as improved legibility), within 18 months of its introduction, the new manager of the ER agreed to let staff stop using the electronic triaging system amidst complaints from nursing staff about the length of time required to triage electronically, which at times left patients at risk.

Making sense of gender, technology, and health sector work

Although the vignettes describe very different technologies and work situations, each tells us something about gender and technology relations in complex organizational settings. In this section, I turn to an analysis of the vignettes. I start by discussing what can be learned about these relations from science and technology studies, before turning to work from gender studies and organizational studies to argue for an approach that combines these different literatures.

Situating technology

Generally speaking, in health care settings, new technology is presumed to lead to better patient care. Some technologies (e.g., medical devices such as x-ray machines and wheelchairs) are subject to rigorous evaluation prior to licensing or use, but such evaluation processes seldom extend to implementation, and stories abound about technologies that failed to work within a specific context, precisely because no one had considered the broad systems into which the technologies would be introduced, and problems came to light only after the new technologies were installed.[4] Furthermore, health information technologies are seldom subject to the same rigorous evaluation standards that are applied to biomedical devices, and problems associated with their use are often viewed by people at a distance from their direct use, as user resistance to change. Part of the reason for this is that technologies are often viewed simply as artefacts—computers, cords, or refrigerators, rather than as complex assemblages of human and technical elements. This was certainly the case in each of the scenarios described in the vignettes. The cord in the drawer (Vignette 1) was initially viewed as a casualty of a move to a new building, rather than a sign that the technical system (keyboards, cords, refrigerators) was not well integrated with the social and organizational systems (good relations with the maintenance staff, clarity in organizational structures about which departments were responsible for which aspects of the machines). In the case of the keyboard trays and the phantom calls, management initially viewed the technical bits (keyboard trays; wireless handsets) as the technology, and presumed that the technologies worked. In both of these cases, problems that arose were initially viewed by hospital management as user ineptness. Problems the nurses experienced with the electronic triage system (Vignette 4) were seen by some as user resistance to change, rather than as a socio-technical system failure. A fairly simplistic view of technology—one which failed to locate artefacts as situated in, and in some sense dependent upon, larger networks of people and technologies in relationships with one another—was implicit in the initial reactions of hospital staff in the first three vignettes. With the exception of the electronic triaging system (where actions taken in response to problems with the system suggested a more nuanced understanding of technology), it was not until our research team contributed to a re-framing of the problems that any resolution to them began to occur.

The interventions our research team engaged in reflected a broader view of technology, which, when applied, rendered the complexity of gender–work–technology relations visible. For the purposes of this discussion and analysis of the vignettes, Bush's early cultural definition of technology[5] is helpful. For Bush,

> technology is a form of human cultural activity that applies the principles of science and mechanics to the solution of problems. It includes the resources, tools, processes, personnel, and systems developed to perform tasks and create immediate particular, and personal and/or competitive advantages in a given ecological, economic and social context.
>
> (Bush, 1981: 1)

This view of technology—consistent with current constructivist views of technology—invites us to view the cord in the drawer (Vignette 1) as something more complex than a problem arising during a move, and the problem with keyboard trays (Vignette 2) and phantom calls (Vignette 3) as something other than a reflection of incompetent technology users. It invites us to see how, in the case of the automated drug dispensing machines, the personnel responsible for fixing cords in place are considered part of the technology of automated drug dispensing, and their involvement (or lack thereof) in addressing the problem falls within the scope of analysis. In the case of the keyboard trays, the staff who were responsible for decisions—from ordering the keyboard trays, to checking that the order was correct when received, to installing the systems and training staff to use those devices—are all considered part of the technological system, and they too would be included in an analysis of problems related to the keyboard trays. Problems outlined with the cord in the drawer and the keyboard trays were both resolved in part as a result of our research team's interventions, which included documenting the problems to a level that they were taken seriously, surfacing problems of accountability within the organization, and introducing a system that supported the management of such technology problems.[6]

Contemporary STS theory such as actor-network theory (ANT) (Law, 1992) suggests that technological systems are the outcome of heterogeneous networks, and that 'society, organisations, agents and machines are all effects generated in patterned networks of diverse (not simply human) materials' (Law, 1992: 2). Following this view, the case of the phantom calls (Vignette 3) could be seen as a failure of the complex network required to enact the wireless call system to properly configure the system. Similarly, in the case of the electronic triage system (Vignette 4), the printers that were supposed to print out completed triage forms (but often failed) would be viewed as part of the technological system of electronic triage, or, in actor-network terminology, would be seen as actors in a network of complex relations, the outcome of which was the series of problems described in Vignette 4.

Clearly the notion of actors enrolled in networks can be used to explain some aspects of how technologies end up in organizations. For example, to realize the wireless call system, several actors, including the vendor, the IT department, the unit manager, the nursing staff, and representatives from the group who maintained the hospital telephone and paging systems, had to come together to form a network in support of that system. In the case of Vignette 4, the manager who made the decision to remove the problematic triage system speculated that one of the things which may have contributed to the system's failure was that the cost of the system necessitated that multiple stakeholders (actors) had to work together to obtain the system, and that each brought a different vision of what the system would do to the project planning process. Each group of actors invested their hopes and dreams in an e-triage system that, in the end, could not meet such varied needs.

Applying theoretical concepts from STS to the vignettes makes it possible to see the problems faced in each context as socio-technical rather than simply technical or user-related. However, concepts from STS do not adequately explain why the dynamics of technological change in organizations often occur along gender lines. Several questions remain to be answered. Was it because the keyboard tray users were women that the concept of ineptness was used to describe users who indicated they were having trouble with their keyboard trays? Was it because the nurses were women that these users were assumed to be making mistakes in assigning rooms to wireless call system handsets? What is the significance of the fact that the electronic triage system had adverse impacts on a largely female-dominated occupation (nursing)? Scholarship from gender studies, particularly work dealing with the relationship between gender and technology, can make an important contribution towards answering these questions.

Women, work, and technology

Each of the scenarios described in the vignettes occurred within the highly gender-segregated world of health care organizations. In all of the vignettes, women were users of technologies but decisions about those technologies (e.g., which electronic triage program to use) were not made by the intended users of the technologies. In all cases, getting the technology to work properly—whether the technology was a cord, a cable, a keyboard tray, a printer, or a piece of software—required interaction with men who were responsible for either maintenance or system design. What is the significance of this nearly universal gendered division of labour for the implementation of IT in health care work? Although early writing about women, work, and technology offers an understanding of how women lose ground relative to men with the introduction of new technology (Bush, 1983; Cockburn, 1983, 1985), the focus in this literature on macro-social forces, such as patriarchy and capitalism, offered few clues about how these social relations are enacted at a micro-level in everyday settings, in multiple organizational contexts, in relation to specific technologies, and at different times and places. Recent work on gender, work, and technology[7] (Webster, 1996; Woodfield, 2000; Herman and Webster, 2007) has built on earlier work which has focussed on issues such as the gendering of skill (Woodfield,

2000), and suggest that 'the gender system of many jobs is in flux' (Webster, 2007: 143), and still subject to negotiation. Harding's early work on the relationship between gender and science (Harding, 1986) offers some insights about the complex ways in which gender relations are reproduced in relation to technology use in organizational contexts in the health sector workplaces from which the four vignettes are drawn.

Harding distinguishes between three aspects of gender: (i) gender structure, or the sexual division of labour (men and women are situated in sex-typed ways); (ii) gender identity or individual gender; and (iii) gender symbolism, a fundamental category within which meaning and value are assigned to everything in the world (Harding, 1986: 57). Harding's notion of gender structure is similar to the notion of the division of labour by gender, in which gender structure is articulated in relation to hierarchical structures of class and race (and, it can be argued, other forms of difference) (Cockburn and Ormrod, 1993). Harding suggested that gender identity or individual gender has two meanings or aspects. Gender identity is projected (potential, actual, or desired identity as others perceive or portray them) and/or subjective (the gendered sense of self—the identity created and experienced by an individual). Gender symbolism involves representations and meanings. For example, when we think of nurses, we tend to think of women, and when we think of maintenance staff and computer support people, we tend to think of men. Gender can be seen as both a relation and a process: masculinity and femininity exist in relation to one another and, while gender gains expression in technology relations, technology acquires meaning in gender relations (Cockburn and Ormrod, 1993).

Applying Harding's gender framework to the vignettes, the initial construction of women as inept technology users who could neither operate keyboard trays nor a wireless call system can be explained in part in relation to Harding's (1986) notion of gender symbolism, a system of representations and meanings which equates technical competencies with masculinity and technical incompetencies with femininity. The observation that the electronic triage system had an adverse impact on a female-dominated segment of the labour force reflects what Harding calls the gender structure—the fact that men and women are situated in different ways in the labour force in general, and the health sector labour force in particular. However, although Harding's framework provides concepts that are useful in explaining how gender works, it falls short of offering an in-depth view of how gender structures, gender symbolism, and individual gender are communicated, enacted, or maintained in organizations. Literature about gender in organizations, and particularly the work of Mills and Tancred (1992) and of Acker (1992), can offer insights into how the mobilization of gender identities and gender symbolism works to maintain gender structures in organizations.

Gendering organizational analysis

Similar to the constructivist turn in technology studies, which saw deterministic views of technology give way to an understanding of technology as socially constructed and the result of negotiations between actors, the 'action perspective' within organizational analysis focuses on how organizational realities are accomplished or negotiated (Mills and Tancred, 1992). Mills and Tancred suggested that organizational analysts should focus on the multiple, interdependent levels and sectors that constitute a given organizational reality, with a focus on process and dynamics. Mills and Tancred (1992) suggested that traditional approaches to organizational analysis are gender blind, and, consequently, considerable errors have been made in interpreting how organizations work. They suggested that it matters a great deal whether or not workers are women or men, and that all of organizational analysis needed to be re-thought on the basis of a gendered substructure. Such an analysis considers both women and men, and places gender at the centre of an explanatory framework. For Mills and Tancred, feminist organizational analyses should be concerned with the contribution of the organization and organizational settings to the maintenance of gender roles. Along similar lines, Acker (1992) pointed out that much early work on women and work, economic and occupational inequality, sex segregation, and so on offered 'no convincing explanations for their persistence or their apparently endless reorganization of gender and permutations of male power' (Acker, 1992: 248–249).

   In each of the vignettes discussed in this chapter, organizational factors contributed to what were often construed as technology or user failures. Curious about why no one had called the maintenance department about the cord in the drawer, the research team queried staff members. We learned that the procedures in place in the maintenance department were incompatible with nursing schedules. Maintenance would not act on a work order unless they could locate the person who had requested assistance. Although most maintenance staff worked days, nursing staff worked in shifts and, often, by the time the maintenance department responded to a problem, the nurse who had reported it was working night shifts or was off duty—between night shifts and her next round of day shifts. Consequently, when maintenance came to address a problem, they would be unable to find someone who could shed light on it, and they would depart from the unit, leaving the problem unaddressed. Nurses quickly learned that taking the time to report a maintenance problem detracted from whatever tasks they immediately faced on the unit and did not result in the resolution of the problem they had reported. This situation offers a good illustration of the point made by Mills and Tancred (1992). Organizations and organizational settings contribute to the maintenance of gender roles. Clearly, such a disregard for nursing schedules—and we could speculate women nurses—could not have persisted had it not been condoned by the organization. The fact that the nurses' complaints about the system were either ignored or viewed as user problems could signal stereotypical attitudes towards women as technologically incompetent.

One of the significant contributors to the situation with the defective keyboard trays was a failure of organizational accountability—itself part of long-established dynamics and enduring organizational processes. Our research team learned that the keyboard tray order was never checked when it arrived, which set the stage for a cascading series of errors. It can be argued that although the absence of organizational accountability was clearly a major contributing factor in the debacle with the keyboard trays, established organizational patterns (where management often insinuated that complaints from staff were frivolous and were indicative of staff resistance to change) initially (incorrectly) identified the source of the problem as the largely female labour force who were regular users of the keyboard trays. Women became scapegoats for a failure in organizational processes. Similarly, in the phantom call case, the problem was initially labelled user error. However, as the situation unfolded, it became clear that the source of the problem was an error that occurred on the part of the vendor and the people with whom he worked to configure the system within the hospital, and not the predominantly female group of nurses who were the largest group of wireless call system users.

All of the vignettes discussed here provide evidence for Mills and Tancred's (1992) claim that it matters whether or not workers are women or men, and their assertion that the organization and organizational settings contribute to gendered social relations. Problems with the electronic triage system, among other things, reflected long-standing relations between doctors (a largely male group in the hospital setting) and nurses (a largely female group). In the case of the electronic triage system, we learned that the doctor who had made the decision to acquire the system asserted that he had consulted nurses; however, nurses felt that they had not been consulted (Balka and Whitehouse, 2007).

Organizations exist within a gendered context, and individuals who together make up an organization do not leave their gender at the door when they go into work each day. Hence, to the extent that technical expertise is generally associated with masculinity (Benston, 1989), within organizational contexts it will be similarly associated with masculinity. Consequently, as was the case with the phantom calls in the wireless call system vignette, although the vendor of the system was an outsider in the organization (which might have suggested he would lack authority within the organizational context), as the only man involved with the wireless call system implementation, he brought his gender— and all the authority associated with masculinity—into the organization and into the encounters described here as the wireless call system implementation. For an extended period, the vendor was able to assert that there was nothing wrong with the technology in spite of numerous problems identified by the (female) nurses.

Through the invisible mechanisms of his gender identity (as a technically competent man) and symbolic aspects of gender (which associated technology with men and caring with women), the vendor entered a gender structure in which all the staff with whom he interacted were women, and which, because of gender symbolism and projected gender identity, pre-supposed his technical competence. He made use of symbolic aspects of gender in suggesting that the users were erring in their interaction with technology and this view overrode the realities of the phantom calls when it became evident that users were not erring in their interaction with the call system. The vendor's prolonged dismissal of the problem of the phantom calls could not have taken place had the others he interacted with during the implementation resisted his conceptualization of the problem as one of user ineptness. Part of the reason the vendor was able to conceptualize the users as inept was because this view matched gendered assumptions and meanings already in circulation in the organization.

For Acker (1992: 250), gender refers to 'patterned, socially produced, distinctions between female and male, feminine and masculine'. Gender, she argues, 'is not something people are, in some inherent sense . . . . rather, for the individual and the collective, it is a daily accomplishment that occurs in the course of participation in work organizations as well as in many other locations and relations'. Acker further argues that 'the daily construction, and sometimes deconstruction, of gender occurs within material and ideological constraints that set the limits of possibility' (1992: 251). Staff come to work with an understanding of gender relations which reflects their lives outside work and, when they arrive at work, their interactions at work further contribute to, or help construct their sense of themselves as gendered people. On a daily basis, the male vendor dismissed complaints from female staff about the phantom calls, during which time he was able to convince some of the unit's managers of his analysis that the problem resulted from users' limitations rather than a configuration problem with the technology.

In discussing gender construction, Acker pointed out that 'the boundaries of sex segregation, themselves continually constructed and reconstructed, limit the actions of particular women and men at particular times' (1992: 251). Relating this to Harding's (1986) concept of gender, it could be said that what Harding refers to as the gender structure influences what actions one can take at any given time. This is an important set of ideas in the context of health sector work. Many women have moved into management positions in the health sector, and increasingly occupy jobs which have historically been viewed as masculine. In addition, power relations between professional groups (such as doctors and nurses) may reflect gendered relations between men and women. Women who cross boundaries (e.g., women who enter management or medicine) may sometimes enjoy the (masculine) status offered by the gender symbolism associated with those professions and may practise in a way that reflects such masculine professional status. This appears to have been the case in relation to the keyboard trays, in which it was women managers who suggested that nurses were inept technology users. In other circumstances, particularly those which involve the traditional technical professions (information technology or maintenance) which Harding (1986) might refer to as having a projected male gender identity, the identity of the profession (information technology or maintenance, which carry a male gender identity) in combination with the fact that it is a male-dominated occupation upon which masculine gender symbolism has been bestowed (when we think of information technology and maintenance workers we tend to think of men) takes precedence over the projected male gender identity associated with (for example) women managers (where being a manager has elements of male gender symbolism), or being a woman doctor (where the projected identity of the profession and the symbolism associated with it project maleness, which can be diminished in strength when the incumbent is female). Put another way, when a woman occupies a position such as doctor or manager where women are in positions typically seen as male, their femaleness diminishes their power in relation to men occupying positions in male-dominated occupations. The sex of an incumbent in a position carries more weight than whether or not the profession is viewed as masculine or feminine.

Greener (2007), who has studied managerial repertoires in the United Kingdom's National Health Service, offers another set of insights which help explain how gendering processes are carried out in complex health care organizations. Greener found that when women occupied senior management positions, management styles (which differed from those of men) emerged in which women appeared to acknowledge that they could not behave 'front stage' (in meetings) in the same informal way that they interacted with others 'backstage' (in one-to-one encounters) or 'off-stage' (in social settings outside of work). Greener observed that both women and men performed differently, depending upon where they were in a public or private setting, and the gender and role (often synonymous with implicit rank) of the person or group they were interacting with. In Greener's terms, gendered impression management strategies are used to maintain face, or, put another way, to preserve symbolic aspects of gender relations in public. As Harding (1986) and others (e.g., Fletcher, 1999) have pointed out, gender is relational, and strategies such as those described by Greener in which management strategies used in one setting differ from those used in another, according, in part, to the sex of those in a group, help to explain phenomenon I have outlined in relation to the vignettes.

In discussing the keyboard tray problem initially (Vignette 2), a small group of female managers came to a consensus that the problem was one of inept users. Functionally, this assertion allowed that group of women to distinguish themselves from the women they managed—in a sense it allowed them to assert the symbolically masculine identity of managers, and enabled them to differentiate themselves from the lower status female administrative staff and nursing staff. When a member of their group eventually accompanied research staff to a unit and experienced a problem with the keyboard trays, this group joined forces and managed a back-stage solution to the problem. Handling the problem backstage allowed them to preserve symbolically appropriate gender relations with the largely male staff and managers of the receiving and maintenance departments (who were arguably responsible for the error), and at the same time maintain the symbolic gender/power relations of managers (symbolically masculine) to administrative staff and nurses (symbolically feminine roles). The women managers therefore preserved the symbolic gender order whilst interacting with their female subordinates and male peers, at the same time that they publicly asserted the symbolically masculine aspects of their jobs as managers within the organizational milieu.

What Acker (1992) has described as the continual construction and reconstruction of gender that limits the actions of particular women and men at particular times can be seen as occurring through, or in relation to, the components of Harding's (1986) gender framework, which are played out in complex ways in relation to technology—and more specifically, technical expertise, in health care organizations. A nurse who becomes an IT specialist operates under one set of gender relations as a nurse (a projected feminine gender identity in a female-dominated occupation), and another set of gender relations as an IT specialist who is part of a management team (both of which have masculine gender identities attached to them). As an IT specialist, and to a lesser extent as a manager, the nurse fills a role in which the gender symbolism and projected gender identity of the role are both masculine. However, if the nurse is also a woman, the strength of that projected identity—and perhaps its currency within the organization—may be weaker than if she was a man. In interactions with female-dominated professions such as nursing, the gender symbolism and projected gender identity associated with being an IT specialist may allow her to reinforce the feminine gender symbolism associated with nursing (e.g., by participating in discussions about nurses as inept users of keyboard trays). But in the presence of a male IT specialist (as was the case with the wireless call system's phantom calls, where the vendor, a man, heavily guarding his technical knowledge), the nurse IT specialist ultimately joined forces with the other women, so that they could collectively mobilize the resources to challenge the male vendor and resolve the problem.

The case of the manager of the ER who decided to allow staff using the electronic triage software to stop using the software amidst problems offers another take on this phenomenon. As both a doctor and a manager, a person inhabits a profession with a very strong projected gender identity associated with both masculinity and technical competence. While being a female (as the new ER manager was) might diminish the strength of the masculine gender identity associated with being a doctor or manager relative to a man in the same role, the strength of both the masculine-projected identity associated with being an ER doctor and the masculine symbolic gender associated with being both a doctor and a manager may have made it possible for the female emergency room doctor/manager to act on behalf of the primarily female nurses in deciding to remove the problematic electronic triage system. Scott and Thurston (2004: 485) have suggested that 'gender is … constituted in a multiplicity of forms through interaction with other symbolic orders, like race and social class'. Presentation and discussion of the vignettes has, I hope, shown how in the work of organizations, gendered rules and roles are enacted in relation to technological change, which, because of gender symbolism associated with both technology and roles such as doctor and nurse, simultaneously reproduces traditional gendered roles for some, while allowing others to redefine gendered roles under particular circumstances.

Discussion and implications

Within the complexity of health care organizations, technology is implicated in gender construction and reconstruction through the work of organizations. As women simultaneously seek to identify themselves with activities such as management, and artefacts such as information technology and, at the same time, navigate complex gender relations within the organization by maintaining symbolically female gender identities in public settings, some aspects of women's stereotypical gender roles (as the inept user) are re-inscribed, while other aspects of traditional gender roles are challenged (the male wireless call system vendor as a technical expert).

Developing a more in-depth understanding of the complex gender dynamics associated with the introduction of information technologies into health sector workplaces could well lead to improvements in IT support in the health sector. As the electronic triaging vignette implies, a greater appreciation of issues arising at the intersection of health sector work, information technology, organizations, and gender could also lead to improvements in the design and implementation of information technologies in the health sector. Accounting for women's work when decisions are made about which system to purchase and how that system should be configured can improve the success of health sector IT projects. Furthermore, to the extent that such health sector technologies produce information that helps us both deliver care and make sense of people's health, attending to these issues can also contribute indirectly to improved health outcomes.

Material covered here may serve as a foundation upon which action aimed at improving practice (here, the design and implementation of information systems in health care organizations) can be built. For example, Harding's (1986) typology of gender can be used to generate questions that could guide practitioners in uncovering hidden gender relations that may threaten the success of IT projects. Understanding the ways that gender, technology, and organizational dynamics work in health care workplaces could support the development of strategies aimed at mitigating gender dynamics at work which interfere with the solution of problems. Although scholarship about women's paid work, gender, and technology has been limited in recent years, the vignettes here suggest that technology continues to contribute to the gendering of women's work, at the same time as it creates opportunities to symbolically escape one's gender identity for some women (nurse IT specialists) under some circumstances (when interacting with other women in equal or subordinate roles). The high costs and low rates of success associated with information technology implementations in the health sector may present a strategic opportunity for researchers and practitioners to intervene in the complex relations of gender, technology, organization, and work in the interest of improving both women's work lives and the chances of success in IT implementation projects.

Notes

1. The lab, located at Simon Fraser University since 1997 under my direction, has held numerous research grants related to information technology in the health sector since its inception. In this chapter, I often use 'we' and 'our' to refer to work undertaken through the lab, in collaboration with students, post-doctoral fellows and others with whom I have worked on projects discussed in this chapter. My use of 'I' reflects my sole authorship of this chapter, and the fact that that views expressed here reflect my own analysis.
2. The first two vignettes are from the Tower Move Project, which occurred between 2003 and 2004, as part of a project titled From Work Practice to Public Policy: Case Studies of the Canadian Health Information Highway, which was funded by the Social Sciences and Humanities Research Council of Canada (SSHRC) between 2000 and 2003, through its standard grants programme. Work on the project which yielded the third vignette began during the Tower Move project, and continued two years later, through the ACTION for Health Project (2003–2008), through SSHRC's Initiative for a New Economy Collaborative Research Initiative funding programme. The fourth vignette resulted from work undertaken through ACTION for Health. For further information about the case studies reported in the vignettes, see Balka and Kahnamoui, 2004; Balka, Wagner, and Jensen, 2005; Balka and Wagner, 2006; Balka and Whitehouse, 2007; Balka, Bjorn, and Wagner, 2008.
3. At the time this situation occurred, I was both a member of a professional practice team at the hospital and a university faculty member with research funding. I used research funding to conduct research about the new technology implementations, which was undertaken in part to support the management team that was responsible for moving staff and equipment into the new building. The project was an action research project in which ethnographic field observations and informal interviews were used to collect data

about issues arising in relation to new technology in the hospital. The project, as well as my team's general approach to action research, is described at greater depth in Balka and Kahnamoui, 2004.

4. A more detailed account of intervention activities undertaken by our team related to Vignettes 1 and 2 can be found in Balka and Kahnamoui (2004).

5. This little-known definition of technology pre-dated much social constructivist work on technological change. I use it here because it is consistent with contemporary social constructivist views of technology, yet invites consideration of power relations which critics of some contemporary theory (e.g., actor-network theory) suggest have been inadequately addressed.

6. See footnote 3 concerning our approach to action research. The first phase of research work undertaken by our team with respect to the electronic triage system is described in Balka and Whitehouse (2007). Subsequent phases of the project are described in publications authored by Pernille Bjørn (Bjørn and Balka, 2007; Bjørn and Rødje, 2008; Bjørn, Burgoyne, Crompton, Hudson, MacDonald, and Pickering, 2008).

7. Little has been written about women's paid work and technology in the last decade. Exceptions are Ruth Woodfield (2000), Joan Greenbaum (2004), Juliet Webster (1996), and work undertaken by Ursula Huws (2007; Huws and Leys, 2003). Huws' work tends to address patterns of change at the macro-level; Greenbaum's (2004) work considers computerization of work over time, in relation to broader managerial forms of control. More recently, a few articles that have addressed technology and paid work have appeared in the journal *Information, Communication & Society*, including the special issue in 2007 on gender (Herman and Webster, 2007).

7

Gendering Work? Women and Technologies in Health Care

Pat Armstrong, Hugh Armstrong, and Karen Messing

All research, as historian E. P. Thompson (1978) said long ago, is a dialogue between theory and evidence. Like geographer David Harvey (2006: 79), we understand theory 'as an evolving structure of argument sensitive to encounters with the complex ways in which social processes are materially embedded in the web of life'. In this chapter, we seek to contribute to that evolution by drawing on our own past dialogues in ways that highlight the central role that gender plays.

More specifically, we build on a wide range of case studies we have worked on over the years to inform theorizing about gender and health information technologies. This is a conceptual chapter in which we explore health information technologies defined as those technologies that support officially sanctioned information impinging on health care work. They encompass the architecture of technologies along with the data they produce and their use. Our space is Canadian health care in the current context of performance measurement, accountability defined in numerical terms, and managerial control aimed at efficiency defined primarily in terms of faster throughput.

Our conceptualization of work combines two linked theoretical approaches. From a determinant of health perspective, all who work in the health care sector are health care workers, including those who provide unpaid care in homes and community settings. This may seem obvious, but increasingly those who cook, serve food, clean, do laundry, clerical and maintenance work are labelled as hospitality workers and defined out of health services. Historically those who do the unpaid care work have been excluded.

From a feminist political economy perspective, women's work is characterized by relations of power and inequality, as well as by integration of their paid and unpaid work (Armstrong and Armstrong, 1990,

2001a, 2004). These relationships have been explored in France by researchers associated with the chair of work psychodynamics and the ergonomics laboratory of the Conservatoire national des arts et métiers. Catherine Teiger (Teiger and Plaisantin, 1984; Teiger and Bernier, 1990) has explained how the demands of women's work in factories and services are poorly understood and undervalued. In Danièle Kergoat's classic treatise on women's work (Kergoat, 1983), she points out that, in the workplace, power relationships follow both class and gender lines. In workplaces governed by a strict hierarchy, women are found towards the bottom, and both minds and bodies suffer from this exploitation. Christophe Dejours (1998b: 135) explains the observation that no one objects to the health damage incurred by low-status workers by the fact that these workers are maintained at a power and perceptual distance from those who determine working conditions. 'Ce n'est pas la rationalité économique qui est cause du travail du mal, mais l'enrôlement progressif de la majorité dans le travail du mal qui recrute l'argument économiciste comme moyen de rationalisation et de justification....' Who benefits is a central question, and contexts—historical, economic, social, and relational—matter.

Our conceptualization of health care starts with these approaches. The work in this sector is highly segregated and female dominated, yet women form a distinct minority of those with decision-making power. This is particularly the case for women from racialized and/or immigrant groups. The work is at least as much about relationships as it is about tasks. It is often associated with what are assumed to be women's innate capacities for caring and with love. It is also often hidden in the household and dismissed as unskilled, and it is rewarded differently for women compared to the much smaller numbers of men who work in care. It is not only the work done by women in health care, but also health information technologies, that are gendered, racialized, and classed.

At least four kinds of information technology are relevant to women's health care work. One sort is the data related to resources such as numbers of hospital beds, employees, and patients, or to the categories of hospital 'separations'. A second form produces and uses data on how work should be done; the protocols, guidelines, task definitions, and care pathways. A third kind of information technology measures and uses data to determine how many employees work for how long, at what speeds, and with whom. Finally, there are the technologies that record the consequences of health care management and practices; the data on such items as return rates and patient outcomes. For ease of discussion, we term these various forms administrative, prescriptive, workload, and outcomes data, respectively.

In exploring the implications of these forms, we understand them as socially constructed in ways that reflect the medical model of health care, with its emphasis on body parts, tasks, and particular forms of evidence. Built into the technologies are assumptions about women, health, skill, evidence, time, and space. These assumptions shape and reinforce certain ideas and practices, ones that typically reflect neither the interests of women in general nor the differences among women. The technologies simultaneously replicate existing power relations and establish new ones, with consequences for women's work and women's health.

This chapter focuses on how these four kinds of information technology are created and employed in health care work, and on their consequences for women as workers. Exploring the development of technologies means raising questions about whose knowledge matters; the kinds of knowledge that matter and the assumptions underlying the development of the technologies. It also means questioning the stated and unstated purposes for which they have been developed, in the light of the uses to which they are put. It means identifying the gendered nature of these purposes and uses, problematizing the dichotomous notions that classify people as sick or well, work as paid or unpaid, and work as a labour of love or not. It includes looking at women's capacity to employ their skills and knowledge; to learn, adapt to, and resist new technologies; to control their work processes; to relate to other workers; and to relate to those using their services. Finally, it means considering the contexts within which health work is undertaken and the ways in which these contexts are taken into account or not.

We begin with an exploration of some of the major forces that shape the architecture, output, and use of these technologies before moving on to consider each of the four in turn.

The contexts for health information technologies

Technologies involve what Lerman, Oldenzeil, and Mohun (2003: 2) call 'the hardening of the social facts'. Health information technologies rely primarily on computers. Computers, as we know them, were initially developed during the Second World War. It was not until 1945 that the name computer became attached to the machine rather than to the women who did complicated calculations by hand (Light, 1999). By the late 1970s, computers had become relatively cheap, small, and easy to use. Based primarily on the rapid calculation of dichotomous choices and a special code language, computer technologies have prior assumptions built into their fundamental structure. They also require a common language, one arising, as all language does, from specific cultural contexts. This means that there are choices built into the very machinery in ways that shape how it can be used, while still leaving some opportunities for users to influence how it is employed. We need to pay attention to both aspects of these technologies; that is, to the assumptions that restrict their use and the possibilities they leave open.

Computer technologies became particularly attractive to health care reformers in the early 1980s for at least two reasons. First, they were presented as reflecting and facilitating access to objective science, and science is understood to be the basis of modern medicine. There is little new about such representation. Technologies have long been promoted as based on science. In dressmaking, for example, the patterns introduced in the late 19th century were represented as having developed from scientific theories about bodies (Gamber, 2003). As with dominant health care practices, dressmaking patterns were designed around the assumption that bodies could be classified into different types and methods developed to fit the figure by means of a set of patterns that could be universally applied. Similarly, the medical model, which still dominates Western health care, assumes that body parts are fixed, and often that these parts respond in similar ways to the invasion of disease regardless of context, differences among groups, and individual response (Armstrong and Armstrong, 2003: Chapter 2).

While this model of the body has contributed to significant improvements in treatment, it has also led to iatrogenic illnesses caused by treatments that fail to address the body as an active, integrated, thinking whole. Similarly, health care work is still planned according to 'scientific management' principles (Gilbreth, 1912) that do not reflect real work activity, its constraints, or individual and collective responses (Armstrong and Armstrong, 2003; Seifert and Messing, 2004). Computer technologies continue to be designed with the assumption that health care is the application of science to universal body parts or universal workers. The technologies allow for the quick transmission of data that has an aura of scientific truth.

The rapid expansion of computer technologies coincided with a renewed emphasis on evidence in health care; indeed, in some ways they made this renewed emphasis possible. Medical researchers argued that too little practice was based on evidence, resulting in both wasted resources and harmful care. What came to be termed evidence-based decision making relied heavily on computer technologies to promote the goal of having the right thing done by the right person at the right time (Sackett et al., 1996). The computer technologies in health care have the advantage of facilitating the rapid dissemination of both information and methods. But, like dressmaking patterns, they also have the disadvantage of suggesting that bodies fit a limited number of de-contextualized categories and that they respond in like fashion. Dressmaking at least assumes difference between male and female bodies, while health care often assumes that scientific knowledge developed from the study of male subjects applies equally to women. As a result, the evidence captured by computers frequently perpetuates assumptions about bodies that deny differences while reinforcing notions of the scientific basis of care.

Second, accessible computer technologies appeared on the market just as health care costs were coming to be described as spiralling out of control. Indeed, all public expenditures on social programmes were similarly represented, even though research suggested that social programmes were not the primary cause of public debt or deficit (Mimoto and Cross, 1991) and that public health care costs were not growing inappropriately (Rachlis, 2005). These technologies seemed to offer a value-free, scientifically based means of managing inputs, processes and outputs in care, in order to increase efficiency and effectiveness. They also seemed to offer a means of ensuring accountability, with accountability frequently understood in terms of the kind of counting that computers can do. At the same time, corporations searching for investment opportunities began to look at the health care sector as a source of profit and to demand reductions in both taxes and overall government intervention in markets. Governments openly embraced and promoted profit-making practices within as well as outside the public sector (Armstrong et al., 2003a). Computer technologies constituted one of those profitable practices. Increasingly, the distinction between public services and private businesses became defined as a problem rather than a virtue. And increasingly, public sector managerial strategies were imported from the private sector, and especially from the Japanese auto industry. 'If you can't measure it, you can't manage it' became a public sector mantra. What could be measured and analysed by computer technologies became the standard for the development, delivery, and assessment of health services and their personnel. Efficiency and effectiveness were defined primarily in financial terms.

While the machines primarily process the information, the computer-adapted language and dichotomous processing limit the possibilities for going beyond these models. Meanwhile, the measurement methods produce data that are treated as facts rather than as the result of particular models of care and specific decisions about what counts as care. Despite the feminization of caring, we have found few examples of attempts to take gender differences into account in constructing the various technologies and little research on their gendered consequences.

Labour accounts for most health care costs. Computer technologies have been increasingly used to reduce these labour costs by providing greater managerial control. These technologies have particular implications for the women who do the overwhelming majority of paid and unpaid health care work. Although the growth in health care costs stems primarily from new drugs and new technologies, monitoring, de-skilling, and reducing the female labour force has become a prime managerial target. The technologies have been used to measure tasks and determine workload, with a view to reducing the time and workers required. Here, too, there are critical assumptions about what is involved in the work. We know from existing research that the skills, effort, responsibilities, and working conditions of women's care work are particularly likely to be rendered invisible and undervalued in such measurement schemes (Steinberg and Haignere, 1987). With limited success, managers in health care have also been trying to have more of the work recorded and delivered by machines. These technologies are usually bought and applied without consulting the women who do the work, despite research showing that the machines force providers to devise ways to reconcile the technology with actual human needs (Teiger and Bernier, 1992; Green et al., 1993; Balka et al., 2006). The resultant application of technology often represents a failure to understand the work required and the particular skills women bring to the job.

Patients, too, have come under scrutiny in efforts to limit their use of the public health care system while promoting their use of private, for-profit care. Computer technologies are developed and employed to ensure just-enough care, with efficiency defined in terms of shorter patient stays and fewer hours with care providers. A clear line is drawn between medical interventions, defined primarily in terms of cuts and chemicals, and care delivered outside health services. This line serves to reinforce a particular medical model that adopts evidence based in large part on male subjects. Women account for a majority of those using health services (Doyal, 1995; Grant et al., 2004). So it is women who are primarily affected by these controls and this model. Equally important, the construction and use of patient databases seldom include a gender, class, or ethnicity perspective that would allow the specific needs of women and men, and of those in different social locations, to be taken into account.

At the same time, the technologies are not used to challenge the gendered segregation of the labour force. Indeed, they reinforce these patterns by building on them. The data are used to send care work home, where it is mainly done by women. Women's responsibility for this growing unpaid care work is required by the failure of the state to provide alternatives (Grant et al., 2004).

Along with the development of these technologies and managerial strategies has come a new language of accountability. In a context of for-profit definitions of success and an emphasis on managerial control, accountability primarily means counting in the ways that computers count. Performance indicators for staff and services are almost exclusively numerical. What is counted, how it is counted, and how easily it can be counted, as well as how the resulting data are used, are determined by managers and their technical services. The emphasis is on speed, control, and cost, with people defined as either sick or well and with success counted as leaving the hospital, but without adequate data on whether they are re-admitted shortly thereafter.

In sum, we argue that contextual social and economic forces help shape health information technologies. The machines and their information do not simply reflect technical expertise or facts. Corporations are interested in selling both their machines and their ways of knowing and doing. They have been supported by governments determined to operate health care like a business; promoting not only for-profit methods but also for-profit services and for-profit attitudes. They have also been supported by a medical model focused on science and evidence of a particular kind. Against the backdrop of these political and economic forces, we now turn to the four specific kinds of health information technology set out in the Introduction.

Administrative data

Health service organizations and governments have long collected data on such matters as number of employees, number of beds, and length of stay. These data have come to be termed 'administrative data', and they have gained increasing importance with the growing emphasis on managerial control and the expanding capacity of technologies to produce data. New areas for counting and control have also emerged along with this capacity. For example, technologies make it possible for governments to pay on the basis of particular diagnoses. This practice reinforces a model of care which treats body parts outside of their social contexts and social relations. As with the other kinds of data under consideration, we touch on one example of administrative data and explore more fully a second example, in order to raise questions about how data are produced and used.

Counting workers

Despite the attention currently paid in health reform circles to the social determinants of health and to teamwork in health care, an increasingly sharp distinction is made between clinical and other workers in health care. The latter are said to perform ancillary or even 'hotel' services, rather than health care services, and often see their laundry, dietary, cleaning, building maintenance, and record keeping work contracted out to for-profit firms. Overwhelmingly performed by women, the feminization of this work leads to its being intensified and corresponding working conditions diminished. In Canada and the United States at least, when contracted out this work ceases to be counted by official statistical agencies as health care work (Armstrong et al., 2007), reinforcing the medical model.

Wait times

The second example of administrative data we consider concerns wait times. In Canada as elsewhere, wait times have become a critical issue, providing a basis for arguments that support the expansion of both for-profit delivery and the private purchase of care. The measurement of wait times thus has profound political consequences. Measuring is complicated by the need to determine when the clock should start, what services or conditions should be assessed, and how appropriate wait times should be determined. What may seem like an adequate response in terms of medical evidence may be experienced as far too long to both the patients concerned about their health and the admission clerks who deal with their resultant anger (Seifert, Messing, and Elabidi, 1999). Moreover, the focus on measuring wait times for some elective surgeries or tests can result in resources being shifted to those services and away from other, equally pressing areas (Carvel, 2006).

A report on wait times commissioned by the Canadian government (Postl, 2006: 61) points out that the 'processes to develop benchmarks, access targets and indicators related to wait times have been concerned primarily with the issues of how to increase efficiency and effectiveness of the health care system to meet these goals'. The report takes up the issue of gender, but only as a result of pressure from the Women and Health Reform Group and then primarily by attaching an appendix, prepared by the Group, to the report (Jackson, Pederson, and Boscoe, 2006). Nevertheless, the report points to a fundamental flaw in wait times data: 'What have (sic) not been addressed in these discussions and research activities is the differential effect that disease, or indeed waiting for care, has on men and women' (Postl, 2006: 61).

The appendix on gender uses the specific example of elective surgery on hips and knees to illustrate the importance of beginning to examine measurement through a gender lens. It starts with the conditions that give rise to different rates and kinds of joint problems for women and men, differences that are about much more than biology. It emphasizes the lives of women, and the ways in which their various household and employment situations have impacts not only on their likelihood of needing replacements, but also on their decision to seek care and start the waiting time clock. It shows the ways in which the care process may differ for women and men as well as among women, and how the impact of outcomes may vary by gender. Class, for example, clearly influences who gets the arthritis that leads to the need for most joint replacements.

These two examples of employee data systems and wait times illustrate just some of the many questions that need to be posed concerning administrative data. The employee data systems deny the contribution of women's work to health care, render it invisible, and reinforce the move to privatize care. The wait times data ignore context and gender, creating spaces for arguments that also promote privatization while failing to provide the kind of information that would be useful in planning for access to care. Similar issues arise when we explore the prescriptive data to which we turn next.

Prescriptive data

Prescriptive data direct the process of care, offering information about how the health care work should be done. Some of these data, such as protocols and medical guidelines, were available and in use before computer-based technologies transformed them. The new technologies speed the calculation and diffusion of such prescriptive data, and thus their implementation, especially under new managerial attitudes. It is now easier, for example, to figure out the most common patterns for diseases and the populations most affected by them. This information can then be quickly translated into policy about treatments and transmitted to practitioners. Other prescriptive data, such as care pathways, directly reflect new approaches to health care that assume it is possible to determine with precision the right thing to do to the right person by the right person in the right time (Berg, 1997). These pathways prescribe the patient journey in some detail, making it possible to remove much of the decision making from providers' hands and put more of it in the hands of managers or insurance companies. In addition, there are technologies that assist providers throughout the care process.

Care pathways

Care pathways set out how long a patient should stay in hospital and what should happen to the patient at each stage of the stay. Such pathways move patients through the system quickly while allowing nurses to follow the pathway without waiting for physicians (Armstrong et al., 2003b; Bourgeault et al., 2004). As a result, this almost exclusively female labour force may gain some autonomy and power in relation to doctors. However, questions need to be asked about the kind of evidence used to develop such pathways in the first place. In the hierarchy of what is considered valid evidence, the randomized clinical trial is at the top, based as it is on notions of single causes and effects of isolated variables. These trials are most commonly conducted on young men, on the assumption that the results are valid for everyone else, despite guidelines developed in some countries to promote the inclusion of women in medical research (Ramasubbu, Gurm, and Litaker, 2001; Hankivsky, 2006). The knowledge that nurses have gained through doing is seldom the basis for care pathways. Yet, because nurses are the ones in daily contact with patients, they have learned a great deal about what does and does not work in practice. They have developed means of responding to particular needs and to particular identifiers of needs through those practices. Instead of incorporating this knowledge, authors of the pathways often seek to prevent nurses from using their experiential knowledge or professional judgement in providing care. As one nurse succinctly put it, 'It gives you a plan to treat the patient that takes away the individual ability to say "You know what? This person's supposed to have learned this by day five and … they haven't got it. So now what's the plan?" ' (Armstrong et al., 2003b: 24).

The knowledge-sharing mechanisms that nurses have developed have also been challenged. Too often, the pathways themselves may be based on actuarial calculations rather than on the limited knowledge that comes from medical research, making cost rather than care the primary criterion for the speed of the pathway. Equally important, such pathways tend to be based on the average patient, too frequently undifferentiated as to gender, age, or culture. They thus reinforce a medical model that ignores social relations and the unequal consequences of faster treatment.

It is also female home health visitors and unpaid female family members who must deal with the inadequate resources allocated for homecare, based on care pathways that often exclude important patient and unpaid provider needs (Grant et al., 2004). Increasingly, hospital workers lack the time to prepare patients and their families to meet their care needs after discharge (Seifert and Messing, 2004). Home health visitors must carry out the screening, coordination, and social support tasks that are not recognized in the organization of their work (Cloutier et al., 1999). Women often take on this work because they have been taught both to do the work and to assume it is their responsibility. But they also take it on because the assessments assume they can and will, leaving them little choice (Armstrong and Armstrong, 2004).

Language services

Language services provide another example of prescriptive data, embodying new notions of how things should be done. In one Canadian case, a hospital serving a diverse immigrant population has contracted out its translation services, which are provided over the phone.[1] The translator is typically located in an overseas country where the language in question is widely spoken, and the patient is located at a nursing station. Not only are sensitive topics addressed in a quite public place, but with a focus on the literal translation of medical terms important issues of social context may be mis-communicated. Not only are women more likely than men to be patients, but immigrant women have fewer opportunities than immigrant men to learn English or French and thus avoid the need to use phone translation services. Thus, the idea that a person from another country, who cannot enter into human contact with the patient and who probably has had no experience of the Canadian health care system, can effectively translate for that patient comes from a highly technical view of what health care is. It seems to us unlikely that the needs of a newly arrived immigrant would be fully served by this procedure. The use of sub-contracted, cheap telephone service is a technical 'fix' involving a truncated representation of health care delivery.

These two examples of care pathways and language services suggest that particular medical assumptions are built into the information technologies. Social relations and other factors that set the contexts for health care and recovery are often ignored or challenged. The knowledge that providers and patients develop from experience is often either ignored or denigrated. These realities are particularly detrimental to women.

Workload data

While prescriptive data addresses how the work is done, workload data are about how many people are needed to do the work and for how long. Here too, assumptions about gender are built into the construction of the data and such assumptions are further reinforced through the ways in which the data are used to determine who works, for how long, and at what jobs (see also Le Jeune, this book).

In their renewed efforts to control the health care labour force and make care more efficient, managers have developed methods to measure the tasks and time involved in providing care. The measurements are in turn used to shift providers around the health services, to ensure 'economic' minimum levels of staffing, and to reduce the overall number of providers. Workload data are also employed to develop performance indicators and to divide up tasks so that more of them can be done by quickly trained workers and/or by contract workers. In this section, we use examples drawn from nursing briefly and from hospital cleaning more fully to pose critical questions about this kind of health information technology. Nurses, 96 per cent of whom are women in Canada, are recognized as professional or at least semi-professional workers, and as critical to care. More cleaners than nurses are men but the work has been highly segregated, with women dominating in what is often termed 'light work' (Messing, Chatigny, and Courville, 1998). In some provinces, a high proportion of cleaners are from racialized and immigrant groups. In order to justify contracting out this work, it is increasingly defined as ancillary, and the specific skills required for cleaning in hospitals are unrecognized. With work defined in this way, it is easier to hire the most vulnerable workers and pay them the lowest wages. However, new managerial practices made possible by new technologies have put both nurses and cleaners under some similar pressures.

Nursing

For nurses, a proposal in the late 1980s to record and measure nursing tasks seemed to offer a way of demonstrating their potential overwork, revealing the complexity of that work and supporting efforts to expand their labour force (Armstrong, Choiniere, and Day, 1993). Canadian nurses have quickly found, however, that the methods adopted for measuring excluded both their own knowledge of the work and critical aspects of their jobs. Bathing provides a classic example, because it involves much more than applying water to skin. As one nurse explained, 'I mean, when a nurse is giving a bath, you're assessing the patient' (Armstrong et al., 2003b: 24); however, it was this type of assessment that disappeared in the time-budget studies on nursing tasks. By reducing care to timed and measured discrete tasks, the measurement system made many of the caring skills disappear along with the notion of holistic care (Armstrong and Armstrong, 2003). Moreover, computer technologies have allowed managers to schedule complex systems by which nurses 'float' around the hospital or even along hospitals, without regard for the valuable contextualized experience that they gain by working in particular areas.

Hospital cleaners

Hospital cleaners too have had their work measured and timed in order to develop precise measures of how long each task should take and how it should be done (Armstrong et al., 1997: 46; Messing, 1998a; Messing, Chatigny, and Courville, 1998). They have experienced even less input to the process than nurses, in part because their work is defined as unskilled and perceived as associated with innate feminine capacities. The intent, as with nursing, is to break down the jobs into discrete tasks. But, similar to nursing work, the cleaners have to deal with both variable patients and variable problems, like vomit in one room, visitors and plants in another, and requests for water in a third. Also, tasks are more than the sum of their operations; there are periods of transition, changes of tools, unforeseen events, interruptions, simultaneous operations, and micro-pauses. Like nurses, cleaners must use their judgement, based on experience and training, to determine when and what work needs to be done. And like nurses, they work as part of a team and in relation to patients. They also face violence, stress, and discrimination that is uncounted. Hospital cleaners have seen their numbers drastically reduced in the wake of technologies used both to measure their work and to determine their work assignments and workload. In fact, the proportion of women employed in hospital cleaning is falling drastically because the tasks furnished to the computer systems excluded or underestimated a number of time-consuming tasks performed more often by women than by men, such as dusting patients' personal items, cleaning dis-used equipment, and polishing mirrors in bathrooms (Messing et al., 1998). This truncated version of women's work has resulted in drastic job loss. According to statistics obtained from the Québec Ministry of Health, the number of 'light work' cleaners (about 90 per cent female) dropped by half between 1994 and 2002, while the number of 'heavy work' cleaners (about 90 per cent male) remained stable. Some of the drop may result in contracting out, which results in deteriorating of working conditions and loss of pay equity wages and job security. In short, women cleaners have felt the brunt of these strategies (Cohen and Cohen, 2005).

Workload data, then, have gender at the core. The way that work is measured and the way the measurement is used to determine workloads reflect and reinforce assumptions about what constitutes care, skill, and women's work. Such data, and their use by managers to determine workloads, have had a profound impact on women's work.

Outcomes data

The final type of information technologies addressed is outcomes data. Outcomes data document the consequences of medical treatment, which are defined as the results of the health care system. For the most part, patients alone are the focus of outcomes data. The emphasis is primarily on death, disease, and direct costs to the system. But changes to health care system have other consequences, such as the shift to community and home care, and altered injury and illness rates for providers. Both affect primarily women.

Community and home care

The shift to community and home care is a shift of responsibility and work to women (Grant et al., 2004). Most of this additional work is unpaid, and research indicates that it often has negative consequences for the health of the women who perform it (Morris, 2001). There are also significant differences in the extent to which women have choices about providing unpaid care, choices related to their financial and other resources. These consequences are rarely treated as outcomes data.

Injury and disease rates

Injury and disease rates are rising among health care providers. Indeed, health care is the most dangerous industry in Canada in terms of workplace-related disease and injury, with rates two and a half times those of the next highest rated industry (Canadian Institute for Health Information, 2002: 87). Needless to say, most of those who are injured and become ill are women. While Canada does collect data on illness and injury for paid workers, they are calculated as a cost of the process rather than as an outcome, with little follow-up on care providers' conditions after they leave the workplace. Data analyses (Lippel, 1999, 2003) suggest that women who are injured or diseased at work have a significantly more difficult time having their disease or injury recognized as work-related. This denial both reflects and reinforces assumptions that women's work is safe and that female complaints are the result of female biology or psychology rather than of women's exposures at work (Messing, 1998b). As a result, even the high, recorded numbers of work-related illness and injuries in health care may understate the problem of gendered health and safety issues at work and its long-term effects (see, Le Jeune, this book, which shows that outcomes data on health and safety costs do not apply well to work done by women).

The point here is to challenge both the boundaries of what is considered an outcome, and whose outcomes are considered. When we take unpaid care work and the long-term consequences of paid work into account, the data on the consequences of reforms look quite different.

Conclusions

This chapter has explored four forms of health information technologies in order to raise questions about their gendered construction and use. Our case study data lead us to conceptualize health information technologies as active tools rather than as passive ingredients, tools that need to be understood within a context of unequal, gendered socio-economic relations. This means beginning with the contexts in which technologies are developed and used, contexts characterized by resource inequalities, gender segregation, and racialization. We must also ask what can be changed in terms of both the architecture and content of information technologies in order to better reflect the interests of women and of different groups of women. We need to assess what cannot be achieved within existing health care technologies but instead requires different methods and means. In addition, we need to ask how these technologies are used, in what context, by whom, and for what purpose. Finally, this chapter has provided some examples that indicate how to think about information technologies in ways that take gender, racialization, and class into account.

Note

1. Sabiha Merali's research on changes to women's health services in Toronto was funded by the Wellesley Institute in a project with Tamara Daly as Principal Investigator. In Canada, only those who move to the country seeking landed status or citizenship are called immigrants. Those working temporarily in the country are called migrants.

8
Ungendering Women's Health: Information Systems and Occupational Health Indicators

Gael Le Jeune

Although women now comprise 47 per cent of the Canadian labour force, they remain concentrated in the service sector (Statistics Canada, 2005). Furthermore, women are more likely than men to experience precarious forms of employment, occupying the majority of part-time, casual, and self-employed forms of work that spread rapidly throughout the labour force in the 1990s (Cranford, Vosko, and Zukewich, 2003). At the same time, women report considerably fewer occupational injuries to Canadian compensation boards than men, accounting for only 30 per cent of compensated injuries between 1997 and 2004, according to the Association of Workers' Compensation Boards of Canada (AWCBC). Women's compensated injury rate may reflect fewer hours spent on wage work (Heisz and LaRochelle-Côté, 2003). Alternatively, it may reflect the different occupations held by women and men. Women are concentrated in a small number of female-dominated occupations that have been compared, perhaps ironically, to the US Food and Drug Administration's 'GRAS' or 'generally recognized as safe' category (McDiarmid and Gucer, 2001: 667). Nonetheless, some authors have challenged this categorization, suggesting that female workers may face systematic discrimination when trying to gain recognition for their diseases, which are often non-traditional (Messing, 1998b; Lippel, 2003). In addition, women as precarious workers may not benefit from the compensation systems to the same extent as their male counterparts (Lippel, 2006b; Bernstein, Lippel, Tucker, and Vosko, 2006). As a result, injuries or diseases prevailing among female workers, such as musculoskeletal disorders or mental disorders, tend to be poorly compensated, and the systems fail to appropriately consider important emerging occupational health problems.

In this chapter, I consider the role played by information systems used by compensation boards in rendering women's occupational health problems in Canada invisible. I explore the role of information systems in the generation of knowledge about women's occupational health through examination of the Quebec compensation board, 'Commission de la santé et de la sécurité du travail' (CSST). The design of such systems is often considered 'gender sensitive', since the sex of claimants is systematically recorded, and compensated injury statistics are usually displayed by sex. However, in the process of constructing compensation injury databases, there is a tragic lack of reflection about the gendered patterns of work and working conditions in Canada that produce very different expositions and health outcomes. I will demonstrate how aspects of filing a claim related to computerization of the claim process deter women from filing claims. This and other factors (that exist in part because existing information systems do not support the collection of data relevant to women's injuries) prohibit us from learning about the extent of women's occupational injuries, which in turn keeps those injuries from being compensable injuries.

Occupational diseases are relatively more prevalent among women (Messing, 1998b), but the information systems in use focus on traumatic injuries (Ison, 2005), with a possible eightfold underrepresentation of certain musculoskeletal disorders, common in women (Stock et al., 2004). Meanwhile, new technologies, such as online claim filling, are implemented without any questioning of the relevance of the information collected from a prevention perspective. Despite this, compensated injuries data are used to produce health indicators which, in turn, influence policies and the allocation of funds and resources. In this context, the investments that are made in information technologies—such as those surrounding the electronic recording of claims—are likely to reinforce the flaws found in existing systems, since they are giving a new materiality to information systems that have contributed to the invisibility of women's occupational health problems (Balka, 2003b). Employing the term 'ungendering', I want to draw attention to the process by which gender issues are made invisible in the practical process of constructing compensated injury databases. In the words of Bowker and Star (1999: 320), 'the moral questions arise when the categories of the powerful become the taken for granted; when policy decisions are layered into inaccessible technological structures; when one group's visibility comes at the expense of another's suffering'. In this chapter, I focus on the taken-for-granted nature of the categories embedded in claims information systems, and show how the information such systems produce contribute to the invisibility of women's occupational health issues.

I begin by presenting a picture of women's occupational health problems drawn from the AWCBC's statistics. I then examine possible gender biases in the construction of compensated injury data following the course of events from the time a woman first becomes sick from her work. Here, I draw on documents (forms, a coding dictionary, annual activity reports, a strategic plan available on line, etc.) and interviews with three key informants to make the role of information systems visible in the larger processes surrounding compensation claims filing, which contribute to the invisibility of women's occupational injuries.

Women's occupational health problems according to work injury, disease, and death statistics

I made a request for the (depersonalized) records of compensated injuries through the AWCBC. This organization compiles compensated injury data from compensation boards across Canada as part of the National Work Injuries Statistics Program, which itself collects data about 'work-related injury, disease or fatality recorded by a compensation board'. AWCBC provided the data for the 1997–2004 period, which I subsequently examined for consistency in coding standards. During this period, the AWCBC recoded almost three million compensated injuries across Canada, including 7000 deaths. Women's compensated injuries comprised 30 per cent of all compensated injuries and only 4 per cent of deaths.

Compensated injuries data are classified into four different categories: accident event or exposure; source of injury; nature of injury; part of body affected. When one examines compensated injuries classified as 'accident event or exposure' (Table 8.1), it can be seen that 27 per cent of injured women are compensated for 'overexertion', the most common type of accident or exposure among women. The second most common accident event or exposure for women is 'bodily reaction' (including allergic or toxic reactions, for example). Strangely, this category is neither an accident nor an exposure. Data about compensated injuries on behalf of men display somewhat similar patterns. 'Overexertion' comprises fewer cases in percentage (22 per cent) and the category 'struck by object' comes second amongst compensated injuries for men.

As for the 'source of injury' (Table 8.2), women are considered to be responsible for their injuries in 23 per cent of cases, compared with 19 per cent of cases for men. 'Floors, walkways, ground surfaces' are

Table 8.1 Principal injuries compensated, by accident event or exposure, Canada, 1997–2004

Principal injuries compensated to women

Accident event or exposure	Percentage ($N = 867,548$)
Overexertion	27%
Bodily reaction	13%
Fall on same level	13%
Struck by object	9%
Repetitive motion	6%
Struck against object	5%

Principal injuries compensated to men

Accident event or exposure	Percentage ($N = 2,070,702$)
Overexertion	22%
Struck by object	15%
Bodily reaction	13%
Fall on same level	7%
Struck against object	7%
Fall to lower level	6%

Note: Only the top six types of accident or exposure have been reported for men and women. Hence, percentages do not total 100 per cent.
Source: Association of Workers' Compensation Boards of Canada (AWCBC).

also important (15 per cent of women's cases). Data about compensated injuries of men display the same patterns, except that, while 'other person' is not an important source of injury for men, it is the third source of injury for women, with 13 per cent of women being injured by someone else.

The 'nature of injury' (Table 8.3) reveals the importance of trauma. Aside from the category 'musculoskeletal system and connective tissue diseases and disorders', which comes fourth, with 7 per cent of cases, the five other principal compensated injury categories are three classes of 'traumatic injuries' and two of 'wounds', representing together almost 80 per cent of cases. As for men, the pattern is the same, except that 'musculoskeletal system and connective tissue diseases and disorders' are less common.

Finally, the 'part of body affected' (Table 8.4) is principally the back (29 per cent of compensated injuries to women). Differences between women and men are few, except women tend to be affected more often at 'multiple body parts' and 'shoulder, including clavicle, scapula'. This

Table 8.2   Principal injuries compensated, by source of injury, Canada, 1997–2004

Principal injuries compensated to women

Source of injury	Percentage ($N = 867,548$)
Person – injured or ill worker	23%
Floors, walkways, ground surfaces	15%
Person – other than injured or ill worker	13%
Containers – nonpressurized	10%
Handtools – nonpowered	3%
Furniture	2%

Principal injuries compensated to men

Source of injury	Percentage ($N = 2,070,702$)
Person – injured or ill worker	19%
Floors, walkways, ground surfaces	11%
Containers – nonpressurized	7%
Building materials – solid elements	7%
Handtools – nonpowered	4%
Not coded	4%

Note: Only the top six sources of injury have been reported for men and women. Hence, percentages do not total 100 per cent.
Source: Association of Workers' Compensation Boards of Canada (AWCBC).

is consistent with other research showing that women are more often found in jobs where the types of manipulations involved result in upper back and shoulder injuries, whereas men are found in jobs where manipulations lead to lower back injuries (Mehlum, Kjuus, Veiersted, and Wergeland, 2006).

The frequencies of traumatic events is striking, as is the fact that mental disorders—evident in qualitative case studies of women's occupational health (Le Jeune, Bélisle, and Messing, 2008)—are excluded or underrepresented in most jurisdictions. Greater differences between men and women's occupational health problems were expected, since they are doing very different kinds of jobs in Canada (Armstrong and Armstrong, 1994; Messing and Stellman, 2006). All of this calls for a better understanding of what is recorded as 'work injury, disease and fatality' and how it is recorded within the different compensation boards of Canada. The Quebec compensation board (CSST) serves as an example through which the invisibility of certain kinds of injuries can be explored.

Table 8.3 Principal injuries compensated, by nature of injury, Canada, 1997–2004

Principal injuries compensated to women

Nature of injury	Percentage (*N* = 867,548)
Traumatic injuries to muscles, tendons, ligaments, joints, etc.	48%
Surface wounds and bruises	12%
Other traumatic injuries and disorders	8%
Musculoskeletal system and connective tissue diseases and disorders	7%
Open wounds	6%
Traumatic injuries to bones, nerves, spinal cord	5%

Principal injuries compensated to men

Nature of injury	Percentage (*N* = 2,070,702)
Traumatic injuries to muscles, tendons, ligaments, joints, etc.	41%
Surface wounds and bruises	15%
Open wounds	11%
Traumatic injuries to bones, nerves, spinal cord	8%
Other traumatic injuries and disorders	7%
Musculoskeletal system and connective tissue diseases and disorders	5%

Note: Only the top six nature of injury have been reported for men and women. Hence, percentages do not total 100 per cent.
Source: Association of Workers' Compensation Boards of Canada (AWCBC).

The construction of compensated injury data

Following the course of events starting with the point at which a woman gets injured or becomes sick from her work, I will shed light on the mechanisms that may explain part of the disappearance of women's occupational health problems from compensated injury data. My insights are based on an examination of documents, as well as interviews.[1] In this section I will show how processes surrounding claims filing present a series of barriers that serve as filters, and lead to an under-reporting of work-related injuries and diseases to compensation boards.

The CSST is a public insurance system responsible for the enforcement of the two main legislative acts that govern the rights and obligations of workers and employers with respect to health and safety in the workplace. These acts are: the Act Respecting Industrial Accidents and

Table 8.4   Principal injuries compensated, by part of body, Canada, 1997–2004

Principal injuries compensated to women

Part of body	Percentage ($N = 867,548$)
Back, including spine, spinal cord	29%
Multiple body parts	9%
Finger(s), fingernail(s)	8%
Shoulder, including clavicle, scapula	8%
Leg(s)	7%
Wrist(s)	6%

Principal injuries compensated to men

Part of body	Percentage ($N = 2,070,702$)
Back, including spine, spinal cord	26%
Finger(s), fingernail(s)	12%
Leg(s)	9%
Shoulder, including clavicle, scapula	6%
Face	6%
Multiple body parts	5%

Note: Only the top six principal injuries compensated for men and women have been reported. Hence percentages do not total 100 per cent.
Source: Association of Workers' Compensation Boards of Canada (AWCBC).

Occupational Diseases (in French 'Loi sur les accidents du travail et les maladies professionnelles' or LATMP); and the Act Respecting Occupational Health and Safety (in French 'Loi sur la santé et la sécurité du travail' or LSST).

On paper, the CSST provides a no-fault insurance coverage for workplace injuries and diseases to 94 per cent of the Quebec workers (Association of Workers' Compensation Boards of Canada, 2005). Domestics, professional athletes, and volunteer workers are not covered. Self-employed workers are not covered either unless they pay into the system voluntarily, which few do; the same is true for domestic workers. It is important to note that most domestics, volunteer workers, and self-employed workers (i.e., most of those excluded from compensation) are women. Even for those who are actually covered, there are many conditions required before a work-related injury, disease, or death can be accepted and recorded as a compensated injury by the CSST.

The first steps in settling the injury claim

The process of compensated injury data construction starts outside the compensation boards with the settling of injury claims. In practical

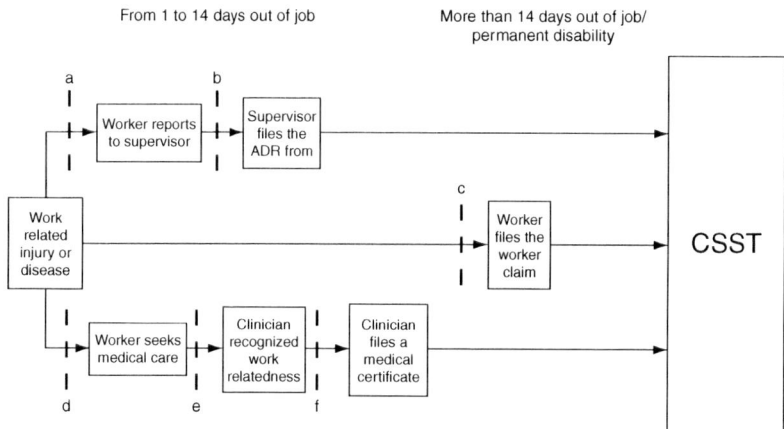

Figure 8.1 Flow chart of events necessary for a work-related injury or disease to be reported to the CSST

terms, a series of barriers (symbolized by dotted lines and labelled with letters on Figure 8.1) may prevent a claim from being reported to the CSST. Webb, Redman, Wilkinson, and Sanson-Fisher (1989), and Azaroff, Levenstein, and Wegman (2002) refer to these barriers as 'filters' that partially explain the under-reporting of work-related injuries and diseases to compensation boards. In Canada, Shannon and Lowe (2002) estimate that workers under-report work-related injuries to compensation boards by some 40 per cent depending on the severity of the injuries. Alamgir, Koehoorn, Ostry, Tompa, and Demers (2006) also describe the underreporting of serious occupational injuries requiring hospitalization in British Columbia.

In Quebec, the ADR form ('Avis de l'employeur et demande de remboursement' or 'Notice of Accident and Claim' in English) is the principal source of information for the compensation board, and this form has to be filled in by the employer. However, some workers may fail to report their health problems to their supervisor (Figure 8.1, label a). Some workers may consider their work-related injury or disease not to be severe enough. Indeed, it is often acknowledged that many workers keep on working even if injured or ill (Brun and Biron, 2006). In fact, some minor injuries or diseases may allow workers to carry on their ordinary activities. Some workers employed in low-wage or 'precarious' employment may also be reluctant to stop working since they cannot afford any loss of income, since they may not benefit from sick leaves, and since they may fear stigmatization or job loss (Azaroff, Levenstein, and Wegman, 2002; Azaroff, Lax, Levenstein, and Wegman, 2004). Some women may well be in this situation. Furthermore, a major reason for not claiming is the availability of other forms of leave. For example, teachers are reluctant to claim for mental health problems relating to stress even though they are compensable in Quebec because of the stressful nature of the CSST process (Lippel, 2006a), and because it has been easier to get paid leave through their sick leave insurance.

Employers may also be reluctant to report work-related injuries or diseases to compensation boards (Figure 8.1, label b) since it may result in a higher experience rating for the employer, increasing the rates they must pay to the compensation board. The CSST actually sets rates according to workplaces' safety performance. This system is supposed to provide a powerful incentive for injury prevention. However, it may also encourage the under-reporting of work-related injuries and diseases (Pransky, Snyder, Dembe, and Himmelstein, 1999). Injured or ill workers' continuance on the job may be encouraged by employers through proposals for modified work. 'Temporary assignment' or 'modified work' has been defined under Quebec law since 1985: section 179 of the LATMP allows employers to assign modified work to employees temporarily, until they are able to return to their regular jobs. Temporary work assignments are not recorded as compensated injuries. Consequently, it seems that employers have increasingly used this procedure as a way to lower their experience ratings since the beginning of the 1990s, when individualized rates were implemented in Quebec (Ouellet, 2003). It seems also that some categories of employers have increasingly resorted to private insurance companies to compensate for long-term absences related to mental health problems, notably in female-dominated professions such as teaching (Vézina and Bourbonnais, 2001; Ordre des psychologues du Québec, 2002). These forms of compensation are not recorded as compensated injuries by the Quebec compensation board either.

After 14 days of absence from work, or in case of a permanent disability or the worker's death, the worker (or her family) is entitled to receive a wage compensation (in French 'Indemnité de remplacement du revenu') which requires that the employer has to pay the worker 90 per cent of his/her net wage for the first 14 days of absence from work, assuming that she (or her family) fills in the Worker's Claim form. However, the procedure may be difficult to understand, and the worker (or her family) may lack the literacy or the support to proceed (Figure 8.1, label c). In Ontario, Finkelstein (1989) has demonstrated that dependents of many men suffering from asbestos-associated diseases and potentially eligible for worker's compensation awards have not received pensions because claims were not filed. Since women's occupational health problems have received considerably less publicity than asbestos-associated diseases, we can expect that a female worker (or her family) will be even less inclined to fill in a Worker's Claim form.

Finally, the worker has to seek medical care (Figure 8.1, label d) since a medical certificate is required. However, the worker may not have any attending physician. She may have particular difficulties in finding a physician who will recognize the work-relatedness of her injury or disease (Figure 8.1, label e) and who will be willing to fill in the medical certificate to be sent to the CSST (Figure 8.1, label f). Gender-role stereotypes and poor training in occupational health may well prevent practitioners from acknowledging the work-relatedness of many women's diseases (Dembe, 1996), but reluctance to get involved in time-consuming, unpaid exchanges with the CSST may also be a factor in physician refusals.

The forms

The forms themselves represent barriers to the reporting of women's health problems since they are primarily designed to report traumatic injuries, more common amongst men than women. The ADR form filled in by the employer and the medical certificate are the principal source of information used to determine compensation. In the process of claim adjudication, additional medical diagnoses may be required, and the CSST may conduct an inquiry at the workplace, invariably if the worker has died. The ADR form provides information about the employee (e.g., name, address, phone number, age, sex, health, and social insurance numbers); the employer (e.g., contact information, number of employees, the CSST code and a prevention mutual code that designates membership in the prevention mutuals that include small and medium enterprises); a description of the task (although here, only the employee's occupation is required), working conditions (which refer to whether the employee is an owner, associate, or administrator; self-worker; working full time; part time; on call; on a seasonal basis; working with a fixed term contract; or other); wages and form of remuneration; information about the 'event' (such as when the event occurred, dates of relapses, recurrences or aggravations); where the event occurred (e.g., at a workstation, elsewhere etc.), names of possible witnesses, and so on.

Theoretically the ADR form provides interesting data, notably about working conditions that may partially explain women's occupational diseases. For example, it has been demonstrated that women paid at piecework rates are more likely to develop disabilities related to cumulative trauma disorders. However, the information about piece-work used to establish that relationship came directly from a joint union-management committee, since the quality of information on completed ADR forms may be uneven. The essential part of the form is a large open field where the employer describes the 'event' with differing degrees of detail. Langley (1995) has summed up the problems encountered with this kind of material: 'There are frequently no guidelines for completing narrative entries, so the consistency which narrative data are entered and level of detail provided can vary by case. In-scope cases may not be identifiable, or circumstances may not be classifiable, due to lack of detail or of specific key words.' It is likely that the CSST will contact the employer or the worker if important information is missing that will affect the decision regarding acceptance or level of indemnity. However, what is relevant to the CSST decision process may differ from what is relevant from an injury prevention perspective. The CSST, like other workers' compensation boards in Canada, sometimes focuses on 'exceptional events' rather than on the 'normal work environment' despite the relevance of the latter in the aetiology of many occupational diseases (Ison, 2005: 17).[2] Consequently, some potentially interesting information may be left out, particularly with respect to women's occupational injuries, since some evidence suggests that they result from the 'normal work environment' to a greater extent than do men's occupational injuries (Messing, 1998b; Le Jeune, Bélisle, and Messing, 2008).

The Worker's Claim form follows the ADR form design. Therefore, it is no better adapted for reporting diseases than was the original ADR form. Since occupational diseases are relatively more prevalent among women than accidental events (Messing, 1998b), the CSST form design actually discriminates against female workers by failing to explicitly request information about the types of injuries women are most likely to experience. Similarly, requesting information about an event reflects the idea that compensated injuries are a result of traumatic events, rather than, for example, exposure over time, a more common cause of women's injuries.

The coding work

According to the former chief of the CSST Statistics Department, adjudicators, the front-line agents empowered to interact with injured workers and their employers, and office clerks are principally responsible for coding claim information. First the eligibility of the claim must be decided on the basis of the completed form and the medical certificate. When a claim is eligible, a file is opened and the 'part of body injured' is recorded by reparation agents. It is unclear who is actually responsible for the coding of information such as age, sex, profession, and information about the working conditions. The coding and recording of the 'nature of injury', 'accident event or exposure', and 'source of injury' are done by office clerks at the closing of the file (i.e., when the worker ceases to be compensated), which could be months or years after the initial injury. Only a grade-five education level is required for these office clerks who are trained in occupational injury and disease classification by the CSST, and there is a high turnover in the occupation. Both factors likely contribute to questionable data quality.

Despite the low level of training required of coders, the classification of occupational injuries and diseases is very challenging. For example, the coders are required to code from incomplete information and to resolve conflicting information (from injury reports and medical documents). To ensure that the coding is 'consistent and accurate', the Canadian Standards Association (2003: 4–5) recommends implementing various measures such as 'checking the appropriateness of selected codes by recoding or reviewing a sample of coded cases' or 'reviewing the rate of NEC/UNS codes, which refer to "not elsewhere classified" and "unspecified" respectively'. However, there are no such systematic quality assessment checks at the CSST, in contrast to practices at the Ontario Workplace Safety and Insurance Board (Van Eerd, Côté, Beaton, Hogg-Johnson, Vidmar, and Kristman, 2006). A single evaluation of the work of the office clerks responsible for data entry was done in 1992 and yielded disappointing results. An epidemiologist interviewed has noticed a systematic increase in missing values from 1995 to 2002 in fields such as 'nature of injury' in the selected parts of the body she is looking at. Overall, in 2002, the percentage of missing values in her data was 8 per cent for injuries and 20 per cent for diseases.

According to the former chief of the Statistics Department, compensation data quality was not a matter of concern at the CSST until recently. The institution has nonetheless undergone profound changes in the work organization of those responsible for data entry. The 2002 CSST activity report indicates that the office clerks of the four administrative areas of the Montreal region were brought together in an operations support centre, which is able to process 60 per cent of the claims and employs only a few. The centre operates by making an early distinction between 'light' and 'heavy' cases. The latter cases represent 10 per cent of compensated injuries but comprise up to 80 per cent of compensation costs because they result in many working days lost or because they become chronic, according to the 2006–2009 CSST Strategic Plan.[3] In fact, the support centre allows human resources to be concentrated on these cases in order to support a fast return to work, referred to as 'maintaining the employment link'. This new work organization also speeds up the processing of claims. Practically, the average time required to ascertain the eligibility of a claim fell from 23 days in 2001 to 13 days in 2005.

A part of the coding work is also vanishing since employers can fill in the ADR form online. New work organization and new technologies have been introduced to save time and money (employers have sent more than 11,000 electronic forms to the institution in 15 months). But there is little evidence of concern about data accuracy—not to mention gender equity—in the innovations promoted, and there is no information available to end users about rejected claims, although statistics pertaining to rejected claims by occupations and sex would be especially relevant to track discrimination against women or other groups of workers. The meaningfulness of such statistics is illustrated in a table presented in the 1931 CAT activity report (Table 1: 18). 'Cases closed where no indemnity has been allowed' are broken down by: an '[inability to work lasted] less than seven days'; a designation of 'no claim'; or 'no accident within the meaning of the Act'; filing claims against an employer not subject to the act, claims disallowed because they were not proved, duplicate claims. Claims could be refused because there was 'not entitled to pension in cases of death', because no notice was given to the employer, or because a claim was withdrawn or not allowed. Unfortunately, this was the only example of statistics that shed light on rejected claims in the 67 CAT and CSST activity reports analysed.

The CSST information system rationale

Primacy of cost control

The CSST's first mandate is to administer the Quebec workplace public insurance system. The organization has to adjust the employers' contributions (their income) to the amount paid out in workers' compensation on a periodic basis. In fact, the concern for financial balance may have increased since the accumulation of deficits in the 1980s and early 1990s (Ouellet, 2003, Figure 4.5: 132). Indeed, in years when the CSST runs a surplus, this fact is inevitably underlined in the CSST President's foreword to the annual report. One of the most obvious rationales for compensated injury data production is therefore to document spending by the system. Interestingly, the CSST's published organizational structure indicates that the Direction of Statistics and Information Management belongs to the Vice-presidency of Finance.[4] The concern for cost control is further corroborated by the interview with the former chief of the CSST Statistics Department. In his words, improving data quality means first of all knowing more accurately where the money goes.

Concern for equity in industry classification

Another CSST basic concern is the grouping of employers into categories of industries with each category assigned its specific premium rate. These rates must reflect the employers' 'experience', that is, the amount of previous compensation paid to their workers. As a consequence, another rationale for data production on compensated injury is the calculation of experience ratings. The CSST is constantly working to improve the perceived equity of the industry groupings and rate setting, under pressure from different industrial groups. The issue of equity in industry classification is almost invariably pointed out in the annual reports. Pressure from employers is obvious since the CSST reasserted its customer service orientation in the early 1990s: In the same vein, one of the CSST slogans in 2000 was, 'Let's allow ourselves to be more influenced by our clients [employers]' (CSST, 2000: 10 translated by the editors).

The focus on industry classification and financial risk assessment in industrial groups may have pernicious effects on workers and seems to free the CSST from assessing occupational risks at the workers' level. Yet, the calculation of financial risks at the employers' level differs basically from the calculation of occupational risks at the workers' level since workers can face quite unequal occupational risks while working in the same industry. For example, the health care sector employs nurses, nurses' aides, and cleaners who display very different patterns of occupational risks. Thus, calculating risk in the entire heath care sector may hide very high risks prevailing in specific occupations within this sector, further masking the nature of women's occupational injury and disease.

The lack of interest in risk assessment in worker's groups

The CSST was originally created to administer the 1979 Act Respecting Occupational Health and Safety. This act seeks to 'eliminate at the source threats to the health, safety and physical integrity of workers'.[5] However, the CSST relies mostly on employers' and workers' observance of safety standards to achieve prevention. A research institute was created (Institut Robert-Sauvé de recherche sur la santé et la sécurité du travail, 'IRSST') but not given a specific mandate for surveillance of rates of injury and illness. Although this has been an important activity of the IRSST since its inception, there is no requirement that this surveillance have consequences for prevention activities other than research. In contrast to many other provinces, the same organization, the CSST, is responsible for both prevention and compensation (Messing, 2002). Its inspection teams are therefore in the paradoxical position of uncovering risks that may justify claims for payment from its compensation arm.

The CSST 2006–2009 Strategic Plan[6] illustrates the misuse of indicators and the confusion that prevails in the identification of hazards. The first strategic orientation is, 'to intervene where severe injuries occur' (p. 4). However, the indicators put forward to achieve this goal are the following:

- variation in number and characteristics of injuries in the sectors concerned by the action plans called 'Construction Industry' and 'Machine Safety';
- number of inspections, remedial orders given, and corrections made, notices of violation in the sectors targeted by the specific action plans, called 'Construction Industry' and 'Machine Safety'.

This extract warrants several comments. First, we may wonder why the CSST is only interested in severe injuries. Less severe ones not only harm larger groups of workers but may also be precursors of more severe damage. In a sense, they are the 'threats to health' the CSST is mandated to eliminate and they warrant attention. Second, it is interesting to learn that the CSST already knows where severe injuries will occur in the next three years. There is no perceived ongoing need to identify high-risk occupations through 2006–2009 injury rates, for example. Third, the variation in number of injuries, considered without a denominator, is a very bad indicator of occupational health risk in a sector. It reflects mainly the variation in number of workers or hours worked in this sector. Also, since women generally work fewer paid hours per week than men, it reflects men's injuries proportionately more on the basis of per hour worked (Messing et al., 1994). In addition, the raw number of injuries is responsive to changes in the composition of the sector about which the CSST does not gather data. For example, if production in the clothing industry is substantially relocated to China in 2006–2009, and this sector suddenly includes a much larger ratio of designers to sewing machine operators, the cumulative trauma disorders common among the sewing machine operators will show a marked decrease. However, it would be wrong to interpret these data as saying that the CSST had been successful in reducing injuries.

In fact, the CSST does not seem to use injury surveillance indicators. This task has been relegated to its associated research centre, and it is unclear how the IRSST research results are taken into account in the CSST strategic orientations. The researcher we interviewed has been working on the identification of target populations and target problems for a long time at the IRSST, usually through compensated injury incidence rates calculated within occupations. He revealed that they have been unable to calculate those rates for recent years (the latest compensated injury incidence rates within occupations were calculated for the 1995–1997 period) because the CSST occupation classification has not followed the evolution of Statistics Canada occupation classifications—in 2001, the National Occupational Classification (NOC) replaced the 1991 Standard Occupational Classification (SOC) at Statistic Canada. Therefore, the numerators (classed by CSST occupation) no longer correspond at all with the Statistics Canada occupations, and there is no alternative means to calculate denominators, for occupations where number of workers or hours worked within occupations are related to the number of compensated injuries within occupations.

Discussion

I have pointed out some problems associated with the compensated injury data construction process which may not be specific to the CSST. Not only is the extent of coverage narrower in the female workforce, but, additionally, even women who are covered may be filtered out in the processes of claim reporting and approval (Lippel, 1999, 2003). Both these processes favour the reporting and compensation of certain types of injury (e.g., traumatic injury, more common amongst men then women), and contribute to the denial of exposures which occur during normal work, which are more common to women workers. As a consequence, the production of compensated injury data does not seem to be gender-neutral. The subsequent lack of data about women's injuries (which reflects the barriers women face in reporting, as well as the structure of existing information systems which do not support the collection of data relevant to women's occupational illness) prohibits us from learning about the extent of women's occupational injuries. This lack of data in turn keeps those injuries from being compensable injuries.

There is no mechanism to keep an eye on the hazards prevailing in vulnerable groups of workers. More generally, there is an absence of reflection about the potential to use compensated injury data for prevention of occupational as opposed to financial risks, and there is some evidence that the indicators used for prevention planning are inappropriate, especially for women (Messing and Boutin, 1997; Hébert, Duguay, and Massicotte, 2003). The confusion and the general lack of coherence may come from conflicting interests of employers who are the 'providers' of the system and workers who are the 'beneficiaries'. The former are bringing money into the system, whereas the latter are spending it; consequently, it is easy to guess whose interests will come first. The lack of coherence may also come from the fragmentation of tasks. Indeed, there is a huge discrepancy between the perceptions of compensation officers, on the one hand, and, on the other hand, those of the office clerks who are actually doing the coding work but whose first concerns may be the speed of claims processing, the CSST central administration, concerned with financial balance, and the IRSST surveillance indicators team, concerned with trends in occupational risks.

Women are put at a disadvantage by the system. Some of them may nonetheless benefit from their employers' private salary insurance coverage and may therefore be able to survive economically, but the compensated injury statistics may be distorted as a result of this transfer from the workers' compensation system to the private insurance system. Thus, women's occupational health problems are underestimated, and the complexities of women's occupational health and safety issues poorly understood. In a vicious circle noted by Messing (1998b), underestimation entails less prevention (Messing and Boutin, 1997) and less research (Messing, 2002), which leads to difficulties in access to compensation (Lippel, 2003).

Despite these problems, there are few other sources of information about occupational health besides compensated injury statistics in Canada. Hence, the AWCBC continues to compile compensated injury statistics from different jurisdictions and tries to harmonize standards. The data appear essential for the association to enhance the safety of workplaces and consequently the health of workers, and the data are made available through numerous tables that can be consulted or downloaded from the association's website. This may be creating an illusion of accuracy and objectivity since numbers now have a sort of 'authority' (Porter, 1995). However, aside from the tables, little is known about the process of construction of the information provided. In fact, with electronic data, the information may travel far beyond the walls of local institutions that produce them, thus obscuring the practical trade-offs that are necessarily made and the power relationships in the process (Bowker and Star, 1999). As a consequence, it is becoming increasingly difficult to make the problems with compensation data visible. What is more, the people who are involved in the production of health indicators may be unaware of what they are actually doing (Balka, 2005). This may be the case with the data stewards at the CSST, who may not be aware of the multiple uses of the numbers they generate, yet they are in most cases the only indicators of occupational health that are being published and used to inform policy.

Acknowledgements

Financial support for this research was provided by the ACTION for Health Project[7] (Professor Ellen Balka, Principal Investigator) as part of the Social Sciences and Humanities Research Council of Canada's 'Initiative for a New Economy' funding programme. I also want to thank Gaston Bernard, Patrice Duguay, and Susan Stock who agreed to be interviewed. I am grateful to the supportive team of the CINBIOSE research centre at University of Quebec in Montreal, where I was located as a postdoctoral fellow. I thank Katherine Lippel for information on the CSST legal context. I owe special thanks to my research director, Karen Messing, for her discerning comments and her translations of some material and to my colleague, Stéphanie Premji, for proofreading this chapter, and Ellen Balka for editorial guidance.

Notes

1. CSST annual reports were reviewed for the years 1980–2005, and the annual activity reports produced by the 'Commission des accidents du travail' (CAT), the institution that was replaced by the CSST in 1980, were analysed before 1980. Forms, the coding dictionary (Canadian Standards Association, 2003), the strategic plan, and so on were examined. I learned that claims are coded according to the Z795 Standard 'which is derived from the US Bureau of Statistics coding standards' (Van Eerd, Côté, Beaton, Hogg-Johnson, Vidmar, and Kristman, 2006: 558). An information manager at the CSST indicated that, to the best of her knowledge, in contrast to the United States (where the history of the coding of occupational injury or disease information has been relatively well documented [Drudi, 1997; Biddle, 1998]), there was no internal document retracing the history of occupational injury and disease classification. Three key informants were interviewed: a retired chief of the CSST Statistics Department, a researcher employed at the Institut Robert-Sauvé de recherche sur la santé et la sécurité du travail (IRSST), the CSST-associated research institute and a physician and epidemiologist working on musculoskeletal disorders

with the CSST data as part of her job as a researcher at the Québec Institut national de santé publique.

2. Ison (2005: 17) explains this tendency as follows:

> The strongest pressure usually operating on workers' compensation systems is the pressure to curtail assessment rates (premiums). Inherently, that pressure is also a pressure to curtail the recognition of occupational disease. Injuries are usually seen to result from exceptional events, whereas diseases often result from the normal work environment. Allowing a disease claim might mean that many similar claims would need to be allowed. So it is usually in disease cases that we find the 'fear of opening the floodgates.' For these reasons, the negative pressure on assessment rates can be much greater in relation to disease cases than injury cases.

3. http://www.csst.qc.ca/NR/rdonlyres/4B4979AB-A22B-4441-97E0-903CBCBD3 7E6/2841/dc_200_2411_2.pdf consulted June 2, 2009.
4. http://www.csst.qc.ca/NR/rdonlyres/A3AFA43F-030C-4831-AF39-1027BCC41 570/5486/Organigramme.pdf consulted June 2, 2009.
5. http://www2.publicationsduquebec.gouv.qc.ca/dynamicSearch/telecharge. php?type=2& file=/S_2_1/S2_1_A.html consulted June 2, 2009.
6. http://www.csst.qc.ca/NR/rdonlyres/4B4979AB-A22B-4441-97E0-903CBCBD3 7E6/2841/dc_200_2411_2.pdf consulted June 2, 2009.
7. http://www.sfu.ca/act4hlth/ consulted June 2, 2009.

9
'It Can See into Your Body': Gender, ICTs and Decision Making about Midlife Women's Health

Eileen Green, Frances Griffiths and Antje Lindenmeyer

Recent research reflects the increasing dependency of medicine on technologies (Webster and Brown, 2004), a form of 'techno-medicine' (Pickstone, 2000) driven by pharmaceutical and medical device industries. However, adopting a sociological perspective suggests that health technologies also need to be understood in social context and as articulated within social relations. Although such technologies provide new and what may be perceived as 'cutting edge' information about disease and illness, they also deliver new forms of risk and uncertainty. Medical imaging technologies in particular can redefine our relationship towards our own bodies and our sense of control over its parts. As Webster argues, 'Technologies perform or "work" within the context of, as well as through, such relationships' (Webster, 2007: 2). This chapter argues that the use of technologies that offer visual representations of the body, paradoxically, tend to both produce and obscure such uncertainties. By appearing to illuminate what has hitherto been an uncertain, invisible risk, the image created becomes an artefact, or one of what Haraway (1990) refers to as 'the crucial tools re-crafting our bodies' (205). In addition such imaging may inadvertently represent normal bodily changes or 'innocent lumps' as risky prior to them (ever) becoming so. This may not be a problem for health professionals who are aware of the complexities of reading medical imaging technologies but when shared with lay women the uncertainties involved may be interpreted differently. We will explore these issues through the examples of mammography screening and follow up ultrasound looking for breast cancer, and bone densitometry, a health technology used to screen for risk of osteoporosis. We characterize mammography and bone densitometry as information technologies because of their potential to provide information about bones and breasts within a technological process, information that is used in health decision making. Interview data from specialist health clinics is presented and analysed in order to illustrate the ways in which women and health professionals attempt to make use of this information when deciding upon the level of health risks posed to individual bodies and subsequent treatment. We argue that analysis of data from the clinics brings into view narratives that equate the technological image with the physical body and in the process, erases medical expertise and everyday interpretative work practices and therefore reinforces the power of the image. Next, we provide a discussion of the gendered dimensions of screening as a background through which to contextualize our research findings from particular breast and bone clinics in the United Kingdom.

The medicalization of women's bodies through screening

The technology used in screening for breast cancer and osteoporosis relates to literature about the medicalization of women's bodies (Lock, 1998; Martin, 1987). Feminists have argued that women's bodies and in particular their reproductive lives (menstruation, childbirth and menopause) have been reconstructed as 'illnesses', a reconstruction that is embedded within a broader process of medicalization, where 'normal events' such as birth, sexuality and ageing are cast within the constantly expanding jurisdiction of medicine. More recent debates critique this thesis as overly simplistic (Harding, 1997), arguing that women are not passive victims of medicine but instead negotiate the range of interventions open to them. An exploration of the socio-technical practices embedded within screening confirms that they are not neutral technological processes; they are also social interactions that include gendered narratives. Gender is implicated in both the narratives that take place in the clinic and the ways in which individual women reflect upon the imperative to participate in screening to reduce health uncertainties. Exploring these narratives will enable us to examine the role played by medical technologies in defining degrees and forms of risk and uncertainty that are embedded within and generated by the screening processes themselves. Howson's work (1998, 2001) has been influential in drawing attention to both the ways in which health screening programmes for prevention are deeply gendered and the absence of gender perspectives within much of the literature on health surveillance and risk. Research on cervical screening has refined the debate about 'surveillance medicine' (Armstrong, 1995), including highlighting the extent to which disease prevention and health surveillance strategies have tended to target women (Graham, 1979). In addition, some feminist work has developed frameworks for exploring issues of power and gender within different types of screening programmes (Bush, 2000; Lupton, 1994; Singleton, 1995). Particularly pertinent are debates about screening as a key process through which women's bodies become subjected to the 'medical gaze' (Howson, 2001). There are now a number of studies of cervical and breast screening that are informed by social science perspectives as well as an expanding body of work on women's experience of the menopause and midlife (Green, Thompson and Griffiths, 2002; Greer, 1991; Griffiths, Green and Tsouroufli, 2005; Lock, 1998; Martin, 1987) but social scientists have paid little attention to osteoporosis scans. Notable exceptions are the work of Guillemin (2000), who examines the significance of bone densitometry and HRT in re-configuring the menopausal body, and Reventlow and colleagues (2002, 2006) who have explored changes in women's sense of embodied identity after bone densitometry.

For both cervical and breast screening in the United Kingdom and elsewhere, national screening programmes backed up by public health material, invitation and reminder letters and health professionals, have succeeded in instilling a strong sense of social obligation into women. Non-attenders for cervical screening are labelled as deviant, since their behaviour departs from the medically constructed norm of having a regular smear test, and they may be blamed for threatening the success of the cervical screening programme (Bush, 2000). In qualitative studies with women in Australia, Lupton (1995) and Willis and Baxter (2003) found that women feel that it is their civic responsibility to ensure that they remain healthy by attending the breast screening programme: 'It is the woman's responsibility to attend for screening, and to attend regularly, or otherwise be held responsible for allowing the unchecked spread of cancer in her body' (Lupton, 1995). Similarly, in our study with women in the United Kingdom, many women saw mammography screening as something they were obliged to do:

> I just think, something that you've got to have done. They send the letter through the door, up to you to just go. I don't necessarily think about it.
>
> (Woman 96)

The gendered nature of the screening process and technology's role in producing seemingly unproblematic measurements both of women's bodies and future risk becomes especially apparent at women's midlife. Lock's (1998) research on managing the post-menopausal body alerts us to the conceptualization of midlife women as situated at one of the 'transitional moments, often associated with danger or crises', a positioning which is implicit in the widespread acceptance of the need for them to engage in medical surveillance such as bone densitometry (Green et al., 2002). Similarly, in the context of screening, risk is seen as objective and measurable, re-enforcing the premise that midlife women's participation in screening programmes is a rational response to risk (Griffiths et al., 2005). Discourses of the menopause as overall decline (Gullette, 2003) are re-enforced by test results that are visualized as fragmented, 'abnormal' or worn-out body parts.

Midlife women as risky body parts

Literature that focuses upon the rise and proliferation of innovative health technologies (Webster, 2002, 2007; Williams, Birke and Bendelow, 2003) provides a useful critique of the implied promise of new health technologies that have become associated with prolonging active life. We argue that it also alerts us to the problem of 'over medicalisation' of bodies in the normal process of ageing. Constructing ageing women as 'abnormal' and a risk to themselves and others encourages them to engage in what Armstrong (1995) and other sociologists (Burrows, Bunton, Muncer and Gillen, 1995; Lupton, 1995) refer to as 'embodied surveillance', which as Howson argues 'heightens and potentially contributes to the development of risk consciousness' (1998: 196). It also signals the need for intervention to reverse or delay such 'decline and deficiency', not least through intervention from medical technologies. Feminist analyses have been critical for opening up debates around menopause but early feminist writing largely neglected to focus on the material technologies of intervention. Also as Guillemin notes, what is missing from most feminist approaches are analyses that explore the relationship between women's bodies, social concerns and the role of technologies in these relations (2001: 200). Theoretical debates about risk and body management have provided useful conceptual frameworks through which to address women's experiences of ageing. Pivotal to our understanding of such experiences are the ways in which women are represented as bodies that need to be carefully managed, whether it be by medical experts or by themselves, through a mixture of personal care and 'self surveillance'. Women's bodies are routinely subjected to screening throughout their adult lives but entering the menopause appears to increase the imperative to take part in surveillance. However, this focus upon the deterioration of specific body parts such as bones and breasts encourages women to experience their bodies as fragmented and increasingly risky with age. Similarly, in producing images that obscure the social context of their construction, the screening technologies themselves become implicated in both a privileging of the technological image over other considerations and the reinforcement of popular perceptions of the image as authoritative and the body as potentially transparent (Treichler, Cartwright and Penley, 1998). Such perceptions obscure the importance of interpretation in the production of health and illness and suggest that we may 'know' our bodies in new and important ways.

The role of ICTs in structuring knowledge about health and illness

Despite popular understandings of the technological imaging that takes place during screening processes as revealing the previously hidden 'inner body', an alternative account locates such imaging within a set of socio-technical practices that themselves produce the body. As Brown and Webster (2004) observe, with the aid of medical technologies, medicine has moved deeper and deeper into body structures. However, it is not only knowledge about bodies that is produced within this process but the body itself that is re-configured with and by technologies. Guillemin argues that 'women's bodies are not neutral sites of biological cell matter' (2000: 193); neither, we would argue, are the technological images produced in breast and bone clinics neutral images that reveal objective knowledge about the state of breasts and bones. Rather, they are highly mediated representations influenced by the location and specific circumstances in which they are produced. As Joyce's (2005) research on MRI scans suggests, the narratives employed in the clinic emphasize the authoritative nature of medical images. Interpretation and discussion of images suggest that they are the key to certainty and knowledge. Furthermore, they infer a view of imaging as providing unmediated access to the body, 'a body that exists outside of human relations and can be known' (Joyce, 2005: 445). Although health practitioners, through their training and work practices, know that such images do not make the body transparent, our data gathered during clinical consultations demonstrates that nevertheless, practitioners use popular phrases in their observations about scan or ultrasound images that obscure this in everyday discussions with patients. Narratives that conflate technologically produced anatomical images with the actual body invited women to believe that such technologies provided unmediated access to a body which can be fully known in ways that were previously not possible.

In the next section we present data from clinical consultations and follow-up interviews with women patients in order to demonstrate that although knowledge in the clinic is co-constructed by health practitioners and women, interpretation of the medical images generated by the screening technologies occupies a central place in both those narratives and what Joyce refers to as the 'tropes' of health professionals (2005: 438). The interpretation of such pictures is dominated by the authority and expertise of the health professional and imbued with the hierarchical status of the clinic where patients are subject to the medical gaze and dependent on the experts to interpret the pictures.

Disappearing 'Spots' in the breast clinic

The data presented below was collected in a breast clinic visited by a number of women as part of our qualitative research project about technologies and midlife women's health.[1] The data set includes a subset of recorded clinical consultations from a breast clinic where ultrasound is undertaken following up 'abnormal' mammography screening results. In addition, we draw upon individual interviews with relevant health practitioners and follow up interviews with women who have attended the clinics. The extracts presented here are from one consultation (Consultation 032) which followed a routine mammogram. The consultation data triangulates with data from an interview with the doctor concerned and a follow-up interview with the same woman a month later.[2] This triangulation allows us to compare extracts from discussion of screening results, types of intervention and treatment, with extracts from the follow-up interviews with the women concerned. The doctor, a female breast specialist, is showing the woman the result of the mammogram repeated just prior to the consultation and compares it to those taken at her routine screening visit. The doctor and patient are both peering at the X-ray image whilst the former attempts to interpret the breast image in lay terms:

> D: And, erm, this is the skin that comes round here, the nipples are here and here in the middle, that fluffy bit is glandular tissue, and the grey and the black bits are sort of fat in between.

D: Erm, and what we do, we just sort of compare each—each half and see that they match. And I don't know if you can see, but you've got a little area (lump) there.

P: That white bit?

D: Yes.

D: That you haven't got on the other side ... ...

D: As you can see, it is very, very small .... I doubt whether there'd be anything to feel at all, but I would like to examine you and do an ultrasound scan, to see if I can, er, spot something. And then I'll talk about the options ... But as you can see, it's—it's—

P: Very small.

D: It's very small, and I don't know if it is anything (to worry about).

P: No. But I'd rather be safe—

D: Of course you would.

During this narrative about the X-ray image, the doctor has indicated that although her 'guess' based upon experience and the size of the 'very, very small' 'white bit' suggested to her that it was nothing to worry about, she wanted to check it further. As the images generated by mammography pick up abnormalities too small to palpate, the superiority of the image over the traditional breast examination is established. What is not open to view, however, is the probable tacit understanding or knowledge held by the doctor that it may also be an artefact produced by the machine itself. The doctor moves between narratives which are designed to re-assure the woman and those which express uncertainty about the nature of the 'white bit'. She concurs with the woman's desire to feel a hundred percent 'safe', with 'of course you would'. However, as a professional, she knows that the imaging technology is not perfect. When interviewed about her general views on screening, she describes her dilemma: she is aware of unnecessary worry and intervention caused by screening but she is 'basically all for it'. She admits that the consultation process leads to obscuring the fallibility of the imaging technology:

I: —they are sent a letter...when they're invited to screening... explaining that it's not a perfect test.[3]

Em, but we don't actually say that to them, em, when we're (pause), pffff, I mean, when—if we find—if we find a—a cancer, that's sort of irrelevant, but if I find, em, well, I mean it's irrelevant in that circumstance— But if I find something on a—on a mammogram, that, em—well, say they come back—they've got a lump and—and we do a mammogram, and the mammogram is absolutely clear, em

I would still examine them and do an ultrasound scan, because I would say to them, em, not every cancer is seen on a mammogram, and some are seen on an ultrasound and vice versa. But I don't point out the—

R: Hmm-hmm.

I: You know, the drawbacks or the benefits of—really, you don't have time. (laughs)

For the doctor, the imperfections of the mammogram are 'irrelevant' if they are beneficial by detecting the cancer; moreover, the availability of another imaging technique, ultrasound, reduces the risk of not spotting a tumour. However, the pauses, repetitions, laughter and spluttering noises as well as lapsing into the impersonal 'we' and 'you' could be indicative of her discomfort in acknowledging the fallibility of the test, and her own complicity in hiding this from the patient. What is also interesting is the relative invisibility of the vital interpretative work and 'professional vision' that is required in reading a mammogram; expertise that is acquired via what Hartswood et al. refer to as 'collective work practices' (2003: 386) engaged in by health professionals within the clinic.

Next in the consultation, the doctor is conducting an ultrasound scan and declares that she cannot find anything [abnormal]. After that, the woman is not, however, reassured.

P: I think I was a bit shocked that you found something or—

D: Right. Now then … this could well be a little blob of—of (normal) breast tissue.

It might be a little tiny cyst that I can't see on the scan; on the other hand, it might be something worrying.

P: Yes.

D: I don't know for certain.

I mean, if I had to guess, I'd say it was towards the least worrying end of—of things—but obviously, we need to know for sure.

P: Right.

D: Now, if you'd had a lump, or something on the scan, I could have taken a sample that way, taken a biopsy—But all we're left with is this (pause) thing.

P: Yes.

D: This little, tiny, tiny area on your (breast)—one view, erm, it's not even on any of the other views. And the only way we could get a—a biopsy, a sample, is by doing what we do a stereo core biopsy.

The immediacy of the image which can be seen by both the woman and the doctor leads to the necessity to act to obtain the best possible certainty that there is no cancer. Both the woman and the doctor are impelled to seek the total certainty that the doctor knows is unrealistic. For the woman, the unfamiliar situation of the clinic, fear about having cancer and submitting to the authority of the expert doctor all contribute to this situation. Being able to actually see the 'blob' on the X-ray without having the expertise to understand the image also contributes to the woman's worry. For the doctor, professional authority is at the same time strengthened (she is experienced in reading the inconclusive images) and undermined (at the end, she remains uncertain and has to resort to more testing).

The outcome of the doctor's uncertainty about the results led to this woman having a core biopsy and waiting 11 days for the result which left her very stressed. Later, during the follow-up individual interview (woman 34), she narrates her feelings about the process and the result, which swing from stating that she was reassured by the good news that the biopsy showed the area was nothing to worry about, to communicating anxiety tied up with the fact that they had seen *something* on the X-ray which although small at the present time, might conceivably become larger and problematic in the future:

I: I mean, as you know, I just had— a breast screening.

R: Yes.

I: And to me, that's absolutely invaluable. Because it was just a routine—mammogram, every three years. And if—if I hadn't had that, I wouldn't have been—(re-called) obviously, I had no problems. They -called me. Fortunately, it's—the news is good. But I think it's so good to have these things. And people are so stupid if they don't use it. Because you think-, I've not had it done for three and a half years. And they found something. Fortunately, it's benign.

The next phrase betrays her anxiety and uncertainty about the result which is mixed in with expressing her strong views on the importance of women complying with regular breast screening. These views are not unrelated to her occupation as a medical receptionist, an occupational identity that appears to make her reluctant to admit to her own anxiety and perhaps even fear about the outcome of the tests:

I: Or they seem to think it's benign. But think, if I hadn't had it done—

> What would have happened? For people like that, in my situation,
> if you didn't have these surveys, you know, this screening rather,
> I mean, a lot of women I know have—refused to have screening,
> refused to have a smear test—Why? They're too scared. What of?
> What have they got to be scared of? It's there, for their health, isn't
> it? And I think that's very, very important, you know, that we have
> this screening.

Later in the same interview she confides how she became very anxious
when kept waiting for the result of the biopsy:

> I: They kept me eleven days waiting, which I thought was extremely
>   poor. The fact there wasn't anybody there (to interpret the biopsy)
>   I was told on the Tuesday that I'd got to wait till the week on Friday,
>   after my core biopsy, for the results. Because eleven days of wait-
>   ing was absolutely appalling. Because at the time the eleventh day
>   came, I went in, and as I say, the—the news was good. But when
>   I came out of (name of building) I just burst into tears.

The woman received the results from a breast surgeon who was not
uniformly reassuring. She described the encounter as follows:

> I: He had no—you know, I'll put you out of your misery. You know,
>   it's—it's clear, you know, the, er, cells appeared to be normal.
>   They've still got the possible marking like a tumour, but, er, we'll
>   recall you in a few months time. You know, and I thought it was a
>   bit abrasive. You know, obviously a tremendous relief. But I couldn't
>   go back to work, at that precise time, because I was too distressed
>   … … … …
> R: I'm glad that the result was good.
> I: Yeah, I think everybody was. You know, it was, uhh, I still have a
>   nagging doubt.

Although the breast surgeon stated that the cell biopsy confirmed normal cells, she picked up that there was something abnormal. It is not
clear whether the 'marking like a tumour' refers to the mammography
or the core biopsy. However, she interpreted the decision to recall her
for repeated tests in a few months time as implying that the finding of
the 'spots' indicated that she might be at future risk. In her interview
it transpires that her doubts relate to hearing about an acquaintance
whose sister had a similar spot on her mammogram which turned out

to be more sinister. The next section of the interview demonstrates two different points, first the importance of informal 'stories' as a source of informal knowledge about health (Henwood et al., 2003) and secondly the powerful impact of the visualization of a benign spot that would not have been detected by palpation or breast awareness. This provides us with a key example of the way in which visualization technologies can themselves construct risk within the body when produced within a clinical context.

> I: …not that they thought there was anything wrong with me. You know, because I didn't even go with a lump or anything, it was a routine, erm, call back, but, erm, somebody I know of, not a friend, she asked me if I was okay, and I says, what had happened, and she says, my sister's exactly the same age as you, exactly—, and the same time as me—had a recall, like me—But she's just had a full mastectomy. And I think, well, I think—and exactly the same thing, with a small mark showing on the mammogram. So I have a little bit of a nagging doubt—that perhaps (pause) erm, they'd missed something. Although they feel it's clear, (and) they recalled me back in a few months. Because of this other woman, who's in exactly the same position—as I was in, hers is full mastectomy. And I think, well, I'm very lucky, but I have that nagging doubt, perhaps.

Her doubts are fanned by the image of the spot portrayed on the X-ray:

> I: —it is something, why is it there anyway? This—why did it show up. Very small—but why did it show up, if it's nothing?

Despite the specialist's reassurance the image of a 'spot' that cannot be explained continues to worry her.

Joyce's (2005) research on MRI scans is useful here in suggesting that the narratives around technological imaging produce and magnify particular ideas. In our case the breast X-ray images are privileged above palpation and breast awareness. This leaves the women described above feeling that the most important result is the image, the fact that 'something' that should not be there appeared in the image of the breast is interpreted as potentially negative. Even though, in our case study, the doctor is aware of the possibility that unexplainable or non-problematic images may appear, the imaging technology is still represented as the agent of knowing, that is superior to other ways of knowing. We could argue that the appearance of 'the spot' actually reveals the instability of the technological imaging process, and required the doctor to draw upon her (subjective) interpretative, professional vision and yet, the presence of the image also impels her to respond by further testing. It is as if currently invisible, risky conditions are bubbling below the surface of the body waiting to be discovered. This concept of the risky body is also apparent in the data from the bone clinic which we discuss in the next section. During the discussions of breast imaging results, health professionals seemed confident in the images' power to detect dangerous changes in the body. However, this certainty was not inherent in one single picture, but in the professional's *interpretation* of different images seen together: X-rays taken from varying perspectives combined with ultrasound scans. This can be seen as an example of a different discourse related to imaging technology, often obscured by the perceived power of the image to speak for itself: that of the health professional's ability, drawn from training and experience, to understand the significance of the image (Joyce, 2005).

All women (and to a lesser degree, men) are of course always at risk of developing breast cancer. While the doctor conducting this consultation insisted that she always told women that breast cancer could occur at any time when giving the all-clear (as she indeed did during another recorded consultation), in some consultations this was not raised. This omission contributed to many of the women interviewed becoming falsely re-assured that having a mammogram every three years protected them for the years in between (Griffiths et al., 2005). This belief may encourage women to relax their personal breast awareness between screenings, relying instead upon the power of the technology to reveal what they cannot know. This devalues women's intimate knowledge of their embodied selves and has implications for early detection of breast cancer by women themselves.

Narratives from the bone clinic

We now turn to the bone clinic where we also find examples of narratives about technology rendering the body transparent. Osteoporosis or 'thinning of the bones' is a highly publicized health condition, mainly associated with ageing women (Gannon, 1999), that is represented as prevented or 'cured' by treatments such as HRT. Testing for 'thinning bones' via bone densitometry has become increasingly demanded by women (post- and pre-menopausal) and is accepted by both women and health practitioners (Richardson, Hassell, Hay and Thomas, 2002) as important in minimizing the risks of painful, disabling fractures associated with osteoporosis. A technology called DEXA scanning produces images of the bone structure using X-rays whilst measuring bone mineral density. The results are presented as a picture, a graph and a table. The graph and table compare the result with the expected result for a person of the same age and sex (based on scanning a sample of the general population).[4] Asymptomatic until a bone fractures, and represented as 'a silent killer', osteoporosis is widely perceived as a women's disease (Richardson et al., 2002) with a projected one in three women likely to experience a fracture during their lifetime (Royal College of Physicians and Bone and Tooth Society of Great Britain, 2001). However, in the United Kingdom there is no surveillance programme as with mammography; women have to be referred to for bone densitometry through their GP.

The presence of the technology and what most women assumed to be precise 'objective' results of the scan seem to heavily influence dialogue in the clinical context. The fact that the key decision making about subsequent treatment takes place most often within a status-ridden, clinical context may make it difficult for women to resist the promise of technology (Brown, Rappert and Webster, 2000). Moreover, women's perceptions of their test result are overshadowed by their fear of osteoporosis which leads them to embrace medical interventions, especially HRT. This fear is fuelled by emotionally charged visual images of the effect of osteoporosis: when interviewed about this topic for our study, many women described their mothers and grandmothers as 'shrunk', 'crippled', 'bent', 'hunched over' or 'crumbled' from osteoporosis.

Similar visual images can be triggered by the image created by bone densitometry technology; also noted by Reventlow et al. (2006) in their qualitative study of women who had undergone a bone scan. As with the breast clinic, data from the bone clinic includes clinical consultations that consider the results of bone densitometry screening, interviews with relevant health professionals and follow-up interviews with women. Our consultation data confirms that clinicians were using bone densitometry screening in a process of risk assessment as part of preventative strategies. However, their women patients were waiting to be told whether they already had or would certainly develop osteoporosis. Their follow-up interviews confirmed that for the women, the clinicians' interpretation of the scan results was a major factor influencing women's decisions about treatment. Interviews with health professionals from across the study provided a different view, with many stating that a low-bone density measurement was only one of several indicators of risk (albeit important). However, being above or below 'normal' loomed large in both the clinical consultations and women's accounts of their consultations at interview. 'Where they are' on the graph became the focus of the consultation.

An important factor in persuading women that bone densitometry was the key to the assessment was their belief that the technology could provide a clear image of thinning bones. The following is from a discussion between a radiographer and a woman while the woman is having a DEXA scan. This provides a vivid example of bone densitometry as a technological artefact, producing pictures of the bone mass body.

> R: Yes it's (scanner) going all the way through and the picture I get on the screen here is, as each time it goes across your body it produces one line of the area that we're scanning.
>
> P: Oh right
>
> R: And it's just a whole load of jigsaws that build up a picture. It shows me how much mineral is in your bones. I'm doing your hip, we know how much you should have for your age, sex, ethnic origin, what is normal for your age, so just compared with the normal data.
>
> P: Oh right, so it's just the mineral content?
>
> R: yes, measuring how much calcium (is) in your bones. (Consultation 49)

The radiographer's role does not extend to interpreting the reading, this is done by a GP or specialist consultant, but the above extract illustrates the important impact which other health professionals' informal interpretations of the scanning process might have, together with their perceptions of the key role played by the scan result in predicting 'normal' or 'at risk' bones. Communicating the result in 'pictures of your bones' appears to help women interpret its meaning but paradoxically also misrepresents what are actually risk calculations, as an individual certainty (Green, Griffiths, Henwood and Wyatt, 2006).

This process is also evident in the following consultation between a woman and a male consultant (Consultation 61). The woman's GP had sent her for a DEXA scan, since an X-ray 'showed some thinning of the bones'.[5] The doctor shows the woman her test results:

> D: Now what the results have shown is this, is the bone scan, is that the bone density, the thickness of the bones is a bit lower than average.
>
> P: Right

D: And just to put it into pictorial context, here is the scan; here you can see the blue band is the normal range.

P: Yes

D: And this little box here represents the bone density measurements of three vertebrae, it's an average result, and as you can see it is a bit lower than average isn't it?

P: Which is the average one? I'm sorry because I should have brought my glasses in with me.

D: The blue band is the average result.

P: Oh right

D: And your result is lower and em ... the scan, can you see this?

P: Yes

D: The scan of the hip shows the same thing that is the normal range and this is your result here.

P: Right

D: So your bones are a little bit thinner and therefore they are more likely to fracture for any given trauma or injury.

What the doctor does not say, but the woman can probably see, is that a 'low average score' is nearer the risk range indicated outside of the blue band. The doctor follows this 'reading off' of population risk for the individual by indicating that treatment may be necessary:

D: What we need to do is find out why they are thin and to advise you about any treatment that might help strengthen them.

P: What makes it ... the bone density go away like that?

D: Well that is what we are trying to find out but.

P: What is the word that I am trying to say ... am I ... I know that I am prone to fractures now because of the osteoporosis em ... I can't think of what I want to say I will just leave it at that.

D: Don't worry about it, it will come out; now tell me have you had any fractures?

P: Yes I have broke a bone in my foot about 3 years ago.

The woman seems to consider that she 'now' has osteoporosis and is therefore prone to fractures. While for health professionals, this likelihood of fractures is a probability, the woman sees it as almost a certainty. This is similar to the perception of the women interviewed by Reventlow and Bang (2006): their body image changed after osteoporosis scans, they saw themselves as in danger of 'breaking apart' and accordingly restricted their movements. Next, the doctor elicits contextual information, asking whether the woman has lost height (she has not) and enquires about other possible risk factors for osteoporosis: age at menopause, smoking, drinking, diet and exercise and family history. After a physical examination, the doctor explains,

> D: Ok now then the reason why you have got thin bones is probably multi-factorial in other words lots of different things may be causing this. One is that it is quite a common condition in women as they get older—there are things that you can do to help.
>
>   And gentle exercise like dancing is quite a good one to and enjoyable to, so you can do that if you took more exercise and if you stopped smoking and if you took a little bit more calcium in your diet that would help you.
>
> P: Yes
>
> D: But you will probably need some drug help as well and the options open to you are things like hormone replacement therapy.

The doctor adds up the risk factors and comes to the conclusion that people with that 'group of things' are more likely to break a bone in the future; however, he seems to disregard the woman's mentioning that neither her mother nor her sister had broken a bone, and that she has not lost any height. In this case, the consultant is exercising his professional vision to digest a variety of clinical information and other data drawn from many sources and applying it to his individual patient. Although it is uncertain whether the woman will have a fracture in the future, the clearest information he can offer is what Edwards and Elwyn (2001) have described as 'subjective probability'. In this case, the doctor chooses to focus upon the factors that contribute to fracture risk, and not those that point to the opposite such as having no relative with osteoporosis. The doctor presents the risk as a certainty which necessitates intervention: 'we know...we must try and prevent'. The consultation continues,

> D: I could see you again in a couple of months' time and see, you know.... There's no great hurry about this, you've had the osteoporosis for a while.
>
> P: How long do you think, (or) you don't know?
>
> D: We don't know really because we've had only one result but it has been there for a while, I'm sure. So there isn't a great hurry to change, but I think it is important that you do take the time to learn up about these various things.

This is the first time that the doctor actually states that the woman has osteoporosis. The image of bone density that is 'a bit lower than average' and therefore nearer to the risk area has been translated into a certain diagnosis which makes treatment necessary. For the health professional involved, the visibility of this measurement straightforwardly relates to the risk of fracture and therefore is an important aid to discussions with the patient:

> I see them as useful aids err to influence the discussion between a patient and doctor err which if you like take a little bit of the wondering out of it, in other words they give a value err or some indication of risk which is a little bit more umm useful than just a clear diagnosis so we give a, a if you like a number to the degree of osteoporosis a person has. So it does enable people to see where they are in the spectrum of other patients' results and prioritize it and make decisions about what they want to do based on an objective measurement.
>
> (Doctor 27)

He refers to the scan result as an 'objective measurement' even though the imaging result cannot stand alone and as Joyce argues in relation to MRI scans, they etch together technology, the body and work practices in complex ways (Joyce, 2005). This trust in health professionals to correctly interpret the results of tests was found in the wider study also. Although the difficulty that health professionals have in interpreting statistical probability has been well described (Gigerenzer, 2002, our data illustrates the power of the 'expert' in a clinical setting. The following extract, from a different consultation, illustrates how the use of data 'triangulation' enabled us to gain deeper insight into the decision-making process. In the clinic, the following exchange occurred:

> P: Do you think that I've got osteoporosis then now, is that what you're saying to me?
> D: You haven't at the moment, as measured.
> P: No, no
> D: But the results are dropping ... You're perfectly fit and well, in good general health. At the moment the results are normal, you have normal bone mineral density, but err after about 10 years it's going to drop into the below normal range, you can't be certain, but it's predictable. (Consultation 01)

In the follow-up interview with the woman, it appears that she had interpreted the 'falling' of the bone density as an indication that she currently had osteoporosis, despite the assurances from the consultant that at that time she was fit and well:

> I do feel that a lot more women should be able to have the bone screening. I do think that would be very important because of the risk of osteoporosis, because I'm probably very lucky that they've found mine.
>
> (Woman 60)

Attending the bone clinic resulted in both women perceiving themselves as having osteoporosis, and therefore more at risk of breaking bones, even though the first only reports experiencing one broken bone and the second was informed that she did not at present have osteoporosis. This intensification of risk perception and internalization of the need for intervention is linked to the power of the visual image. The perceived implications of bone mineral density becoming 'lower than normal' lead to women embracing HRT even when the results are currently normal, and dropping below normal levels is only a prediction for the future.

Conclusion

The presence of the technology and what most women assumed to be precise 'objective' results heavily influenced the dialogue in the clinical context. The computer-generated images of bones provided by the DEXA scan and breast 'spots' visualized by mammograms can be shared with women in the process of interpretation by clinicians. If the image is interpreted as being a direct representation of the body, what it shows cannot be ignored: the strange 'spot' or the scan image was there, the woman has seen it; therefore, it cannot be 'nothing'. Due to time constraints and professional hierarchies and protocols, interpretation work is seldom made sufficiently visible in the clinical consultation process, which encourages a privileging of the technological image over other considerations and led to these women becoming convinced that they were being informed about certainty of breast cancer or bone fractures. Following Callon and Latour (1981) and Ormrod (1995), Joyce (2005) uses the term 'black-boxing' for the tendency of imaging technologies to obscure the practices surrounding them. In the consultations discussed here, images 'black-boxed' the uncertainty inherent in any risk calculation, and the imperfections of the imaging process itself. Moreover, they helped to obscure the role of the health professional in interpreting and translating the image; turning 'spots' and a little box on a blue line into a recognizable picture of the woman's body. This led to ongoing anxiety about a possible tumour for the woman attending the breast clinic, and to certainty of now having osteoporosis and therefore being prone to fractures in the future for women in the bone clinic.

The gendered discourse of the female body as 'risky' was influential in these consultations, even in the breast clinic where the consulting specialists were aware of the shortcomings of imaging technologies and highly reliant upon their collective professional expertise to interpret any images correctly. Women developed a new view of the body as fragmented, an object of observation (Brown and Webster, 2004; S. Reventlow, 2002), and a source of anxiety or danger. Both the representation of midlife women's bodies as risky and in need of surveillance—which is reinforced by public health information—and the immediacy of the visual image compel the women to see themselves as 'at risk' and to accept, maybe even demand treatment. Although machines cannot, of course, speak for themselves, the interaction of the seductive powers of imaging technologies and representations of the female body as in need of surveillance meant that women saw their bodies as sources of increased disease risk, while at the same time desiring certainty that they did not have breast cancer, or that they would not become stooped old ladies. Health professionals, while aware of the impossibility of these wishes, colluded with them by not challenging this desire. Most recent research highlights increasing debate about the potential of screening to do harm by increasing anxiety and unnecessary invasive treatment (Gøtzsche and Nielsen, 2006), challenging the widely accepted 'it's better to go and know' mantra. As our examples illustrate, a conflation of material bodies and technological images, alongside gendered discourse of risky female bodies can combine to construct a certainty of harm, with the attendant potential to reduce women's confidence in what may be normal ageing bodies.

Notes

1. The research study *Innovative Health Technologies at Women's Midlife: Theory and Diversity between Women and 'Experts'* was funded by the Economic and Social Research Council (grant number L218252038). The study report can be accessed in full from the ESRC's website http:// www.esrcsocietytoday.ac.uk/ESRCInfoCentre. 109 consultations with midlife

women were recorded, both in primary care (p. 72) and in secondary care (p. 37). Of these, 29 were in clinics focusing on the interpretation of bone densitometry and 8 in a breast assessment clinic. Thirty-seven patients in these consultations and 58 health professionals gave a follow-up interview; additionally, 61 women recruited from the community were interviewed.

2. All data from the study is available for bona fide use via the Economic and Social Data Service Qualidata: http://www.esds.ac.uk/qualidata/(accessed 19 June 2006). Participant codes in this article follow those submitted to the archive.

3. This refers to the leaflet 'Breast screening—the facts' that every woman is sent with her invitation letter to attend breast screening (see Department of Health, 2007 for the latest edition). The leaflet warns women that breast screening can miss cancers; that it does not prevent cancer, and that women can develop cancer in between their mammograms.

4. The bone mass measurement recommended by the World Health Organization in defining osteoporosis is 2.5 standard deviations or more below the mean bone density for premenopausal women (Wilkin, Devendra, Dequeker and Luyten, 2001).

5. Reading the whole of this consultation reveals some confusion. She is hearing impaired. She relates much of the consultation to her foot: she suffers from Sudeck's atrophy since she had a fracture three years ago.

Conclusion: Re-configuring the Gender, Health and Technology Relationship

Eileen Green, Flis Henwood, and Ellen Balka

Our aim in this concluding chapter is to reflect upon the relationship between gender, health, and ICTs, paying particular attention to the social contexts within which they become embedded. We return to the two key questions posed in the Introduction: How does exploring the socio-technical relations of ICTs in health care inform our understanding of gender? And to what extent can a closer understanding of gender as process, or of how gender 'works', inform and extend our knowledge of the ways in which ICTs are being developed, implemented, and used in health care contexts? This final chapter seeks to reflect on the ideas developed by individual authors in order to return to the overarching themes addressed in the book, to consider their implications for policy and practice, and, finally, to pinpoint future directions for research.

All of the chapters in this collection draw upon empirically grounded studies of ICTs in health care. The contributors share a commitment to understanding the significance of the settings or domains in which the technologies under investigation are developed and/or used, and to understanding how gender both shapes, and is shaped by, the socio-technical changes that they have identified in their empirical observations. Taken together, these contributions enable readers to begin to develop a more nuanced understanding of how technology contributes to the development and maintenance of gendered relations and practices, including: how gendered relations are implicated in the uptake and use of technologies, and how technologies, in their capacity as infrastructure, impact upon the structure of knowledge about gender and ways of knowing about health and bodies. In the Introduction to this book, we suggested that there were three broad cross-cutting themes in this collection. These were identified as focusing on health information seeking and gendered care-giving in informal contexts,

ICTs and gendered paid work in formal health care settings, and ICTs and gendered ways of knowing about health. We shall now discuss the book's contribution in each of these areas.

Health information seeking, informal contexts, and gendered care giving

A number of chapters in the book have explored the ways in which gender and ICTs are intertwined in the processes and contexts of informal care giving. Key themes include: how the technology is used in the process of health information seeking in informal settings, how gender mediates the use of the Internet to access health information, the emergence of the 'informed patient', and the role of information intermediaries in the process of informal care giving. At the beginning of the book, Henwood and Wyatt's chapter argues that the Internet has become a key actor in the informed patient discourse and that both the dynamics surrounding Internet use and those of health maintenance are gendered. Their analysis suggests that contextual factors such as age differences and gender relations within households mediate use and that both health and the Internet are sites of gender construction, involving processes and practices that construct informed patient discourses. Although such health and Internet practices can signal changes in everyday behaviours (e.g., increased information seeking), at the same time, other practices and discourses, including those related to gender, can remain resistant to change or be re-inscribed. Simpson, Hall, and Leggett raise similar issues in their chapter, which focuses on how gender mediates the uptake and use of the Internet in rural and remote settings. They suggest that health is a complex, personal issue that is situated firmly in the social, rather than the technological realm. Although the presumed efficiencies of ICTs for providing health information to widespread consumers is important, Simpson et al. argue that the key role of informal, gendered social support networks in ensuring delivery of/access to ICT-based services is too often ignored by government-led health programmes. Like Henwood and Wyatt, they highlight the critical role of health information intermediaries in empowering health care support practices that are always both contextual and characterized by gendered social relations.

Simpson et al.'s chapter also highlights people-centred health strategies by describing the ways that information seekers obtain assistance from significant others in contextualizing information. Individuals turn to other people as sources of health information, mainly because they are accessible and trustworthy, and can assist with information appraisal, as well as providing emotional and informational support. In their current format (one which predominantly follows a model of information broadcast, rather than dialogue or information exchange), health information websites do not meet the needs of many health information seekers because they fail to provide the full range of support mechanisms which health information seekers desire. The research findings of Simpson et al. in rural areas of Australia include observations of health information intermediaries using both symbolically masculine and feminine skills as they carry out their work. This chapter raises questions about whether representations of caring work as feminine (low-tech, hands-on, and emotional work) have contributed to an under-resourcing of vital health intermediary roles.

The obscuring and undervaluing of unpaid caring work is a key theme addressed by other authors in this collection, including Harris, who explores gender relations in the care work performed by librarians engaged in delivery of ICT-based health information services. Harris argues that libraries play a key role in the provision of access to health information but, because their mainly female practitioners are not perceived to be ICT experts, the personal support and intermediation provided by them is concealed behind their representation as anachronistic, an image that belies their key support for citizens searching for health information. Informal care work is also addressed in the chapter by Bella, who, like Henwood and Wyatt and Simpson et al., analyses 'gender work' that challenges traditional understandings of masculine or feminine work. Presenting empirical data from her case study of support for public access computers at a remote community centre, Bella argues that the gendered caring work performed by both men and women in this context is more complex and nuanced than traditional representations of caring work suggest. In the quest to empower users at the MacMorran Community Centre, paid workers and others moved between nuanced degrees of masculine and feminine caring: organization-based executive caring, emotional caring about, and 'being cared for', in their use of ICTs.

In addition, both Bella and Simpson et al.'s chapters demonstrate that, in the context of informal work, the role expectations related to health and technology embedded in everyday practices may also challenge traditional gender practices. Although chapters by Balka and Sharman address the interaction of gender, technology, and health in women's *paid* labour, their research findings yielded similar insights. Despite popular representations of particular occupations and work practices as stereotypically gendered (nursing and caring as feminine, IT and management as masculine), such practices often work, implicitly, to continuously negotiate and renegotiate what Harding (1986) refers to as our projected gender identities (how other's perceive us in gendered terms) and in doing so, our relationships with technology. Gendered behaviours and social practices influence how we interact with co-workers and/or family members and inform our identification with, and emotions about, everyday caring activities (e.g., feeling that we *should* look after our sick Aunt Junie), and form part of the contexts within which women interact with health sector ICTs. Such behaviours and emotions also inform our interactions with technologies, at times reinforcing those stereotypically gendered behaviours and on other occasions, often via interaction with specific technologies, challenging them.

This grouping of chapters demonstrates how the use of ICTs in both formal and informal health care settings can blur distinctions between paid and unpaid work practices (e.g., those of a caregiver and a librarian), between different kinds of paid and unpaid caring work, and in some cases, between gendered behaviours. For example, Harris described the dilemmas librarians faced when, as a result of the presence of Internet terminals in public libraries, people newly diagnosed with serious illnesses sought not only support in conducting information searches, but also social support and help in coping with their illnesses. Similarly, Bella's depiction of the multifaceted caring relations surrounding support for ICT use in a low-income community highlights the complexities of gender relations represented in caring work that becomes embedded in ICT support practices. We need to continue to ask questions about the specific role of ICTs in these contexts, and the ways in which gender becomes differentially inscribed in the social and work practices represented in a variety of informal care contexts.

Formal health care settings, ICTs, and gendered paid work

Several of the chapters have addressed the complex interactions of gender and technology in formal health care settings. Although the health contexts described by Sharman and Balka in their chapters differed markedly, the strategies that they observed health sector workers were employing as they navigated gender–technology–organizational relationships were similar. Balka's chapter focused on the implementation of several different technologies in varied clinical settings and on the social relations between front-line staff and managers. In contrast,

Sharman's chapter focused upon emergency room nurses, who were required to define their job content in relation to administrative work, which they viewed as lower status work than nursing. In spite of these differences, staff interacting with information technology in both settings drew upon varied but similarly gendered strategies in their representation of what Harding (1986) might describe as symbolically feminine aspects of their work environments (e.g., caring aspects of their jobs for nurses in the emergency department, and appropriately 'feminine' service behaviours when on front stage, for female managers). Interestingly, in both settings, technology-related practices were implicated in the ways that professional staff—whether female nurses or women managers—distanced themselves from lesser skilled women with whom they worked. Hence, both chapters have illustrated the complexities of the ways in which implementing new ICTs in health sector workplaces can lead to changes in work practice for women workers whilst simultaneously reproducing hierarchical gendered identities.

The theme of gendered patterns of job segregation in health-related occupations explored by Sharman and Balka is continued in the chapter by Armstrong, Armstrong, and Messing. Drawing upon case-study data amassed over a period of years, Armstrong et al. demonstrate that different kinds of information technology are created and used in health care work to collect numerical data about both administrative matters (counting beds, length of patients' hospital stays, etc.) and employee data in the health service, which redefines the type of jobs that are included as health related. The information systems used in such change processes reclassify occupations routinely perceived as women's work (cleaning, laundry, and dietary services) as non-clinical, which allows them to be contracted out to for-profit organizations, whereupon such work ceases to be counted by official statistical agencies as health care work, becoming de-valued and more poorly paid. Armstrong et al. assert that assumptions about gendered work are embedded in the technologies that shape and re-enforce ideas and practices that routinely fail to reflect the interests of, and differences among, women.

ICTs and gendered ways of knowing about health

The remaining chapters offer another perspective on gender and ICTs in health care. The indirect influence of ICTs on the construction of gender is the subject of Le Jeune's chapter. In her analysis of occupational health and safety and related databases, it becomes clear that our understanding of gender in relation to occupational health and safety issues is constructed to a significant degree by database design and architecture, which influences which data are collected and which are omitted. Our understanding of gender issues in the work place are constructed in part by silences or omissions: by what we do *not* know about women's health at work. Le Jeune's research demonstrates that information system design contributes to such silences through her exploration of the ways in which online claim filing systems reduce the quality of data that might be used to evidence the need to compensate women workers suffering from occupational health problems. The theme of gaps or omissions is continued in Green, Griffiths, and Lindenmeyers' chapter which shows us how new technologies are implicated in the construction of normal ageing processes as a disease, presenting evidence which suggests that patients and providers alike seem to ascribe authority and certainty to the results of screening technology, to a degree which is unwarranted. Green et al. present data about the social practices embedded in the use of imaging technologies from breast and bone clinics, in support of their argument that such visual imaging technologies are represented as 'agents of knowing' within contexts that are both gendered and hierarchical. The reduction of women's bodies to fragmented parts illuminated upon screens encourages women and health professionals to engage in narratives that imply a capacity for such technologies to reveal risky and hidden parts of the body. The clinical focus upon the reading of such images 'etches together' aspects of the body with political hierarchies and technical practices that 'produce' the body (Grosz, 1994). This both valorizes the place of visual images in the consultation process and obscures the importance of professional and lay interpretation of these and other symptoms, encouraging women to devalue their everyday knowledge of their own bodies. In this way, technologies used in clinical health care contexts indirectly influence the construction of gendered roles, and gendered selves.

Taken together, the chapters demonstrate the varied ways in which gender–technology relations shape our attitudes about health as well as behaviours that contribute to health, sometimes reinforcing and at other times challenging projected and symbolic aspects of gender. Focusing on how gender works, and particularly how it works in relation to technology use undertaken to support health, simultaneously highlights the possibility for change in gender–technology relations related to health, and serves as a reminder of the weight of gender as a symbolic and projected device in relation to health and technology.

Implications for policy and practice

Health policies in industrialized countries continue to promote the notion that information and communications technologies improve health care—whether through increased provision of consumer health information in support of self-care, or through the implementation of widespread health information systems that can network within and between health provider units and organizations. Improved service delivery and organization is often understood as 'a given', following on from the application of new technologies, without due attention being given to the social and organizational changes necessary to support these overall health service improvements. This book has drawn attention to the *gendered* social and organizational relations that can inhibit both the successful implementation of ICTs in health care organizations and the types of individual and collective empowerment often seen as accompanying ICT applications in the field of consumer and public health.

In their different ways, contributors to this book have shown how the very nature of the gendering process, with its emphasis on *difference* at the symbolic level and *divisions* at the material level, has meant that practices associated with femininity (if not always solely with women) are often ignored or undervalued in decision making about ICT implementation and use. Such oversights have resulted in a number of different kinds of problems, often affecting both those who seek to implement new systems and those who seek to use them. The research discussed in this book has shown how this gendering of ICTs in health care has resulted in problems of user resistance to new technologies (Sharman chapter; Henwood and Wyatt chapter); inappropriate resourcing for the caring and support work necessary for making Internet-mediated information useful for the specific individuals and communities (Harris chapter; Bella chapter; Simpson et al. chapter); poor design of technologies and information systems that reproduce social inequalities (Le Jeune chapter; Armstrong et al. chapter) and problems of governance and management of newly implemented systems (Balka chapter) which often result in inefficient use and disillusioned users.

There are implications for both policy and practice-based stakeholders here. In countries such as Canada, the United Kingdom, and Australia (as covered in this book), which have used state funds to underwrite the cost of computerization in the health sector, governments now need to invest, equally, in the human resources necessary to support the effective utilization of these new information systems. This investment must be on the basis both of an improved understanding of existing work practices (including the gendered divisions of both formal and informal labour that accompany them) and a detailed plan for how work will be organized and managed within the new model of care being promoted. Wathen et al. (2008: 191) in their discussion of the policy and practice implications of their research on health 'info(r)mediaries' make a similar point, arguing that any reallocation of resources in favour of training for human intermediaries must start with an acknowledgement of the importance of such work. There are significant roles here for professional associations and workers' representatives (such as trades unions) who must advocate for their members, both by drawing attention to how new technologies may reproduce existing inequalities and by negotiating for a recognition of the skills and competences necessary for more effective uses of these technologies. Such recognition may open up new career and promotional opportunities for women and other marginalized workers and carers.

Most chapters in this book are based on research that reports the implications of the gendering of health care information technologies, with many making a plea for greater understanding of the ways in which such gendering processes contribute to the further invisibilisation of women's work (or, in some cases, of feminine work practices, as performed by both women and men). However, central to the research reported by Balka in her chapter was an action-oriented research approach that may provide a useful model for how practice-based interventions might be conducted in the future. In this model, researchers are both analysts and practitioners, seeking to better understand *and to improve* the processes of ICT implementation in specific health care settings and contexts. Such approaches may not be suitable in all research contexts but it seems to us that where such opportunities do present themselves, they should be taken. Research that seeks to influence technological change in a more progressive direction might even be welcomed by governments and health care organizations who are increasingly concerned about user resistance and the seemingly unreachable goals of their technologically driven, more cost-efficient and effective delivery models.

There is also a much bigger role for public participation in the debate about how ICTs are being implemented and used in health care. Several chapters in this book (Green et al., Le Jeune, Armstrong et al.) have drawn attention to the ways in which gendered assumptions are built into the very design of health information technologies which, in turn, reinforce gendered practices and even gendered 'ways of knowing' about health. These observations draw attention to the very social character of technologies that is often obscured in dominant discourse that presents such technologies as neutral and yet authoritative. For example, Green et al. showed clearly that, despite the uncertainty and ambiguity surrounding their use, the 'visual' health information technologies involved in screening programmes carry a weight and authority that often fuels public demand for greater access to such technologies. More concerted efforts to engage the public in debates about the technologies and the wider social contexts of their design and use could result in more appropriate levels of uptake as well as more careful and critical readings of the 'results' such technologies produce. Such public engagement or 'involvement' is now encouraged in health care decision making but rarely do notions of public engagement extend to include a focus specifically on the health care information technologies themselves. Were they to do so, the social (including gender) dynamics of technological change could be fore-grounded. Such fore-grounding would need to take place within a framework that understood technologies not as tools, systems, and devices for improved health care organization and delivery but as part of a wider 'sociotechnical ensemble' (Bijker, 1995) that, if designed to incorporate context-sensitive configurations of people and technologies, could constitute a truly innovative, user-centred, and democratic model for health care delivery.

The case for more interdisciplinary research?

In the opening chapter of this book, we suggested that three key bodies of literature—the 'gender and technology', the 'technology and health', and the 'gender and health' literatures—have shaped our approach to the book. Following chapters have demonstrated how attention to such literatures has shaped particular projects and, in turn, how the findings and interpretations that emerge from these projects have contributed to existing debates. We now return to these central themes, suggesting ways forward for future research in the field. The strength of this collection is that it draws together the work of researchers from varied disciplinary and practitioner backgrounds that have adopted a variety of theoretical and methodological positions and carried out their research in very different contexts. The collection is also innovative in offering insights into the three-way intersection of gender, technology, and health, in ways that have not previously been attempted, exploring their co-construction across a broad range of settings and contexts.

What should future research priorities be? We address this question by returning to the gender and technology debate, reviewed in the introductory chapter of this book. Over the years, a large body of theoretical and empirical work on gender and technology has emerged with valuable contributions from feminists (e.g., Wajcman, 2004, 2007; Wyatt, 2008). This work has had some notable impacts, including: enhancing the social shaping of technology debate through developing understandings of the mutual shaping of gender and technology, framing debates on gendered identities and technology, and opening up the gendered dimensions of ICT work, access, and use. Turning our attention to the theorization of gender (relations) within health-related settings has offered new insights and suggested new directions for future research within the gender and technology field.

Most recently feminists (Green and Singleton, forthcoming) have begun to explore the lack of attention to gendered social relations within new theorizations about technologies badged under the label 'Web 2.0' (Beer and Burrows, 2007), technologies which are already having a big impact within health arenas (Hardey, 2008; Eysenbach, 2008). In post-industrial, post-production, consumer-based Western societies, new digital technologies are playing an important role in shaping contemporary social relations, and in (re)configurations of identity and community and modes of sociality. Social shaping theory (SST) has shown us that technologies are socially contextualized and that their design, development, utilization, domestication, and rejection are shaped by contemporary social relations (Silverstone and Hirsch, 1992; MacKenzie and Wajcman, 1999). Yet, as argued throughout this book, crucially and often overlooked, these relations are also gendered. The fact that the social processes inherent in technological development, uptake, and use are gendered (Green and Adam, 1998) has significant implications for the shaping and reshaping of gendered identity and gender in lived social relations (Adkins, 2004). It is through the type of micro-social analysis that is evidenced in the empirical research described within this collection that we make visible the evidence of 'doing' gender, that is, gender as process. We need to focus upon the everyday (Green, 2001), in this case within health arenas, in order to understand the relationship between processes of technological innovation and the ways in which various ICTs are consumed and become domesticated. Rather than concentrating upon the potential of the digital to achieve the extraordinary, we need to remember that it is the capacity of ICTs to become a routine part of, and constitutive of, 'normal everyday life' which is important.

Empirical research on gender and technology (Green and Singleton, 2009) can unpeel the layers of complexity in everyday life and consistently reveal those aspects of gender that are transformed and those that endure in new socio-technical contexts. As contributions to this book have shown, the workplace, health, and caring, all key foci for feminists in earlier decades, remain as important as ever. More research which links health, gender, and ICTs will both enhance our understanding of how gender 'works' within different health contexts and reveal the ways in which it becomes (re)inscribed within ICT-related health care practices.

We are aware that we have posed more questions than can be answered at this point, but are convinced that it is an optimum time for more nuanced empirical research in health contexts which study the intersections between gender and ICTs. As argued above, there is an increased awareness that for ICTs to be user friendly and designed as fit for purpose requires fine-tuned and detailed knowledge of the contexts and practices that do and will shape such technologies. Equally, as feminists continue to debate (Balka, this book, Faulkner, 2001; Lie, 2003; Wajcman, 2007), gender relations are constantly in process and, like ICTs, evolving and changing over time, which signals the necessity for drawing more users (women and men) into the design circle. This is especially important in health arenas since they are increasingly characterized by digital devices set to transform both health care systems and practices and the functioning and representation of physical bodies (consumers and providers alike). It is important that we support more interdisciplinary research in order to capture the intertwining of gender, technology, and health as complex and mutually shaping processes across time and diverse spaces. This collection has moved us forward in such a quest; however, there are of course gaps and omissions. By facilitating a diversity of theoretical approach, this book both represents and has sought to further explore, competing understandings of, and approaches to, the relationships between gender, technology, and health. As such, it can claim to be both a useful introduction for scholars seeking to enter this field of study and a timely reminder to more experienced researchers of the contested nature of knowledge in this emergent and interdisciplinary field.

Bibliography

Acker, J. (1992) Gendering organizational theory, *in* A. J. Mills and P. Tancred (eds) *Gendering Organizational Analysis*. London, Sage, 248–259.

Adam, A., Emms, J., Green, E. and Owen, J. (eds) (1994) *Women, Work and Computerization: Breaking Old Boundaries-Building New Forms*. Amsterdam, Elsevier.

Adkins, L. (2004) Introduction: Feminism, Bourdieu and after, *in* L. Adkins and B. Skeggs (eds) *Feminism after Bourdieu*. Cambridge, Blackwell, 3–18.

Alamgir, H., Koehoorn, M., Ostry, A., Tompa, E. and Demers, P. A. (2006) How many work-related injuries requiring hospitalization in British Columbia are claimed for workers' compensation? *American Journal of Industrial Medicine* 49: 443–451.

Armstrong, D. (1995) The rise of surveillance medicine. *Sociology of Health and Illness* 17: 393–404.

Armstrong, P. and Armstrong, H. (1990) *Theorizing Women's Work*. Toronto, ON, Garamond Press.

Armstrong, P. and Armstrong, H. (1978) *The Double Ghetto: Canadian Women and Their Segregated Work*. Toronto, ON, McClelland and Stewart.

Armstrong, P. and Armstrong, H. (1994) *The Double Ghetto: Canadian Women and Their Segregated Work*. Toronto, ON, McClelland and Stewart.

Armstrong, P. and Armstrong, H. (2001a) *The Double Ghetto: Canadian Women and Their Segregated Work*, 3rd edn. Toronto, ON, Oxford University Press.

Armstrong, P. and Armstrong, H. (2001b) The context for health care reform in Canada, *in* P. Armstrong, C. Amaratunga, J. Bernier, K. Grant, A. Pederson and K. Wilson (eds) *Exposing Privatisation: Women and Health Care Reform in Canada*. Aurora, ON, Garamond Press, 11–48.

Armstrong, P. and Armstrong, H. (2002) *Planning for Care: Approaches to Health Human Resource Policy and Planning*. Ottawa, ON: Commission on the Future of Health Care in Canada, Discussion Paper No. 28.

Armstrong, P. and Armstrong, H. (2003) *Wasting Away: The Undermining of Canadian Health Care*, 2nd edn. Toronto, ON, Oxford University Press.

Armstrong, P. and Armstrong, H. (2004) Thinking it through: Women, work and caring in the new millennium, *in* K. R. Grant, C. Amaratunga, P. Armstrong, M. Boscoe, A. Pederson and K. Wilson (eds) *Caring for/Caring About: Women, Home Care and Unpaid Caregiving*. Aurora, ON, Garamond Press, 5–43.

Armstrong, P., Armstrong, H. and Laxer, K. (2007) Doubtful data: Why paradigms matter in counting the health-care labour force, *in* V. Shalla and W. Clement (eds) *Work in Tumultuous Times: Critical Perspectives*. Montreal, QC and Kingston, ON, McGill-Queens University Press, 326–348.

Armstrong, P., Armstrong, H. and Scott-Dixon, K. (2006) *Critical to Care: Women and Ancillary Work in Health Care*. Toronto, ON, National Coordinating Group on Health Care Reform and Women/National Network on Environments and Women's Health.

Armstrong, P., Armstrong, H., Bourgeault, I., Choiniere, J., Myhalovskiy, E. and White, J. (2003a) Market principles, business practices and health care: Comparing the U.S. and Canadian experiences. *International Journal of Canadian Studies* 28: 13–38.

Armstrong, P., Armstrong, H., Bourgeault, I., Choiniere, J., Myhalovskiy, E. and White, J. (2003b) Space, place and time: Managing nursing care in Canada and the United States. Paper presented at the Third Biennial Gender, Work and Organization Conference, Keele University, UK.

Armstrong, P., Armstrong, H., Choiniere, J., Myhalovskiy, E. and White, J. (1997) *Medical Alert: New Work Organizations in Health Care*. Toronto, ON, Garamond Press.

Armstrong, P., Choiniere, J. and Day, E. (1993) *Vital Signs: Nursing in Transition*. Toronto, ON, Garamond Press.

Association of Workers' Compensation Boards of Canada (2005) *2004 Key Statistical Measures*. Retrieved 31 January 2008 from: http://www.awcbc. org/common/assets/ksms/2004ksms.pdf.

Aune, M. (1996) The computer in everyday life: Patterns of domestication of a new technology, *in* M. Lie and K. Sørensen (eds) *Making Technology Our Own, Domesticating Technology into Everyday Life*. Oslo, Scandinavian University Press, 91–120.

Azaroff, L. S., Lax, M. B., Levenstein, C. and Wegman, D. H. (2004) Wounding the messenger: The new economy makes occupational health indicators too good to be true. *International Journal of Health Services* 34: 2271–2303.

Azaroff, L. S., Levenstein, C. and Wegman, D. H. (2002) Occupational injury and illness surveillance: Conceptual filters explain underreporting. *American Journal of Public Health* 92: 1421–1429.

Azuma, K. (1990) The relationship between nursing status and sex-role. *Sangyo-Soshiki Shinrigaku Kenkyu* 4: 3–16.

Babrow, A. S., Kasch, C. R. and Ford, L. A. (1998) The many meanings of *uncertainty* in illness: Toward a systematic accounting. *Health Communication* 10: 1–23.

Baines, C. T., Evans, P. and Neysmith, S. M. (eds) (1998) *Women's Caring: Feminist Perspectives on Social Welfare*, 2nd edn. Toronto, ON, Oxford University Press.

Baines, C., Evans, P. and Neysmith, S. (eds) (1991) *Women's Caring: Feminist Perspectives on Social Welfare*. Toronto, ON, Canada, McClelland and Stewart.

Bakardjieva, M. (2005) *Internet Society: The Internet in Everyday Life*. London, Sage.

Bakardjieva, M. and Smith, R. (2001) The Internet in everyday life. *New Media and Society* 3(1): 91–107.

Balka, E. (1997) *Computer Networking: Spinsters on the Web*. Ottawa, ON, Canadian Research Institute for the Advancement of Women.

Balka, E. (2003a) Getting the big picture: The macro-politics of information system development (and failure) in a Canadian hospital. *Methods of Information in Medicine* 42: 324–330.

Balka, E. (2003b) The role of technology in making gender count on the health information highway. *Atlantis* 27(2): 1–13.

Balka, E. (2005) The production of health indicators as computer supported cooperative work: Reflections on the multiple roles of electronic health records, *in* E. Balka and I. Wagner (eds) *Reconfiguring Healthcare: Issues in Computer Supported Cooperative Work in Healthcare Environments: Workshop Proceedings ECSCW 2005*. Burnaby, BC, ATIC Design Lab, 67–75.

Balka, E., Bjorn, P. and Wagner, I. (2008) Steps toward a typology for health infor-matics. *Computer Supported Cooperative Work (CSCW 08)* (pp. 515–524). San Diego, November 8–12.

Balka, E. and Kahnamoui, N. (2004) Technology trouble? Talk to us! Findings from an ethnographic field study. *Proceedings of the Eighth Conference on Participatory Design (PDC '04): Artful Integration: Interweaving Media, Materials and Practices Vol. 1*. New York, ACM Press, 224–234.

Balka, E., Kahnamoui, N. and Nutland, K. (2007) Who's in charge of patient safety? Work practice, work processes and utopian views of automatic drug dispensing systems. *International Journal of Medical Informatics* 76: S35–S47.

Balka, E., Messing, K. and Armstrong, P. (2006) Indicators for all: Including occupational health in indicators for a sustainable health care system. *Policy and Practice in Health and Safety* 4: 45–61.

Balka, E. and Smith, R. (2000) *Women, Work and Computerization: Charting a Course to the Future*. Boston, MA, Kluwer Academic Publishers.

Balka, E. and Wagner, I. (2006) Making things work: Dimensions of configurability as appropriation work. *Computer Supported Cooperative Work Conference 2006 (CSCW'06)* (pp. 229–238). Banff, Alberta, November 4–8.

Balka, E., Wagner, I. and Jensen, C. B. (2005) Reconfiguring critical computing in an era of configurability, *in* O. W. Bertelsen, N. O. Bouvin, P. G. Krough and M. Kyng (eds) *Critical Computing: Between Sense and Sensibility. Proceedings of the Forth Decennial Aarhus Conference* (pp. 79–88). Aarhus, Denmark, August 20–24. New York: ACM Press.

Balka, E. and Whitehouse, S. (2007) Whose work practice? Situating an electronic triage system within a complex system, *in* E. Coiera, J. I. Westbrook, J. L. Callen and J. Aarts (eds) *Information Technology in Health Care 2007: Proceedings of the Third International Conference on Information Technology in Health Care. Socio-Technical Approaches Vol. 130 Studies in Health Technology and Informatics*. Amsterdam, IOS Press, 59–74.

Ball, M. J., Hannah, K. J., Newbold, S. K. and Douglas, J. V. (eds) (2000) *Nursing Informatics: Where Caring and Technology Meet*, 3rd edn. New York, Springer.

Banerjee, I. and Leong His-Shi, C. (2006) Internet in the war against HIV/AIDS in Asia, *in* M. Murero and R. E. Rice (eds) *The Internet and Health Care: Theory, Research and Practice*. Mahwah, NJ, Lawrence Erlbaum, 357–374.

Banks, I. (2001) No man's land: Men, illness and the NHS. *British Medical Journal* 323: 1058–1060.

Barefoot, J. C., Gronbaek, M., Jensen, G., Schnohr, P. and Prescott, E. (2005) Social network diversity and risks of ischemic heart disease and total mortality: Findings from the Copenhagen City Heart Study. *American Journal of Epidemiology* 161: 960–967.

Beauvoir, S. (1952) *The Second Sex*. New York, Bantam.

Beer, D. and Burrows, R. (2007) Sociology and, of and in Web 2.0: Some initial considerations. *Sociological Research Online* 12(5). Retrieved 11 November 2008 from: http://www.socresonline.org.uk/12/5/17.html.

Bella, L. (1995) Gender and occupational closure in social work, *in* P. Taylor and C. Daly (eds) *Gender Dilemmas in Social Work, Issues Affecting Women in the Profession*. Toronto, ON, Canadian Scholar's Press, 107–124.

Bella, L. and Bishop, R. (2004, October) *Community Capacity Development: A Framework for Understanding the Contribution of Community Access Computers at MacMorran Community Centre to Community Capacity, Literacy and Health.* Paper presented at the Canadian Public Health Agency Conference on Literacy and Health, Ottawa, ON.

Bella, L., Harris, R., Chavez, D., Fear, J. and Gill, P. (2008) 'Everybody's talking at me': Situating the client in the info(r)mediary work of the health professions, *in* C. N. Wathen, S. Wyatt and R. Harris (eds) *Mediating Health Information: The Go-Betweens in a Changing Socio-Technical Landscape.* Basingstoke, Palgrave Macmillan, 18–37.

Bella, L. and Kearley, M. (2005, May) *Re-Imagining Community Development: The Role of Community Access Computers in the Development of Individual and Community Capacity.* Paper presented at the meeting of the CASSW, Congress of the Humanities and Social Sciences, Winnipeg, MB.

Bella, L., McDonald, G. and Walsh, R. (2004, October) *Capacity Development and the Urban CAP Programme at MacMorran Community Centre.* Paper presented at the Canadian Public Health Agency Conference on Literacy and Health, Ottawa, ON.

Benston, M. (1989) Feminism and systems design: Questions of control, *in* W. Tomm (ed.) *The Effects of Feminist Approaches on Research Methodologies.* Calgary, AB, University of Calgary, 205–224.

Bercovitz, K. L. (1998) Canada's active living policy: A critical analysis. *Health Promotion International* 13: 319–328.

Berg, I. (1997) Problems and promises of the protocol. *Social Science & Medicine* 44: 1081–1088.

Berg, M. (2004) *Health Information Management: Integrating Information Technology in Health Care Work.* London, Routledge.

Bernstein, S., Lippel, K., Tucker, E. and Vosko, L. F. (2006) Precarious employment and the law's flaws: Identifying regulatory failure and securing effective protection for workers, *in* L. F. Vosko (ed.) *Precarious Employment: Understanding Labour Market Insecurity in Canada.* Montreal and Kingston, McGill-Queen's University Press, 203–220.

Beynon-Davies, P. (1999) Human error and information systems failure: The case of the London ambulance service computer-aided despatch system project. *Interacting with Computers* 11: 699–720.

Biddle, E. (1998) Development and application of an occupational injury and illness classification system, *in* S. D. Stellman (Chapter ed.) *Encyclopaedia of Occupational Health and Safety, 4th edn, Part IV: Tools and Approaches, Chapter 32: Record Systems and Surveillance.* Retrieved 8 June 2009 from: http://www.ilocis.org/en/contilo.html.

Biddulph, M. and Blake, S. (2001) *Moving Goalposts: Setting a Training Agenda for Sexual Health Work with Boys and Young Men.* London, Family Planning Association.

Bijker, W. (1995) *Of Bicycles, Bakelite and Bulbs: Towards a Theory of Sociotechnical Change.* Boston, MIT Press.

Bijker, W., Hughes, T. and Pinch, T. (1987) *The Social Construction of Technological Systems.* Cambridge, MIT Press.

Bjørn, P. and Balka, E. (2007) Health care categories have politics too: Unpacking the managerial agendas of electronic triage systems, *in* L. Bannon, I. Wagner,

C. Gutwin, R. Harper and K. Schmidt (eds) *ECSCW 2007: Proceedings of the Tenth European Conference on Computer Supported Cooperative Work.* London, Springer, 371–390.

Bjørn, P., Burgoyne, S., Crompton, V., Hudson, D., MacDonald, T. and Pickering, B. (2008) *Boundary Factors and Contextual Contingencies: Configuring Electronic Templates for Health Care Professionals.* Manuscript submitted for publication.

Bjørn, P. and Rødje, K. (2008) Triage drift: A workplace study in a pediatric emergency department. *Computer Supported Cooperative Work* 17(4): 395–419.

Bourgeault, I. L., Lindsay, S., Mykhalovskiy, E., Armstrong, P., Armstrong, H., Choiniere, J., et al. (2004) At first you will not succeed: Negotiating for care in the context of health reform. *Research in the Sociology of Health Care* 22: 263–278.

Bowker, G. C. and Star, S. L. (1999) *Sorting Things Out: Classification and Its Consequences.* Cambridge, MA, MIT Press.

Braidotti, R. (1996, July 3) Cyberfeminism with a Difference. Retrieved 25 January 2008 from: http://www.let.uu.nl/womens_studies/rosi/cyberfem.htm.

Brennan, P. F. and Fink, S. V. (1997) Health promotion, social support and computer networks, *in* R. L. Street, Jr., W. R. Gold and T. Manning (eds) *Health Promotion and Interactive Technology: Theoretical Applications and Future Directions.* Mahwah, NJ, Lawrence Erlbaum Associates, 157–169.

Brown, J. S. and Duguid, P. (2002) *The Social Life of Information.* Boston MA, Harvard Business School Press.

Brown, N. and Webster, A. (2004) *New Technologies and Society: Reordering Life.* Cambridge, Polity Press.

Brown, N., Rappert, B. and Webster, A. (eds) (2000) *Contested Futures: A Sociology of Prospective Techno-Science.* Aldershot, Ashgate.

Brun, J. P. and Biron, C. (2006) Absentéisme et présentéisme: Entre la maladie, la paresse ouvrière et la responsabilité professionnelle. *Actes du Colloque—La Recherche en SST: Anciens Risques et Enjeux Actuels.* Congrès de l'ACFAS (17–19 mai, Montréal, Canada). Retrieved 31 January 2008 from Réseau de recherche en santé et en sécurité du travail du Québec: http://www.rrsstq.qc.ca/stock/fra/fichier0182.pdf.

Bunton, R. and Crawshaw, P. (2002) Risk, ritual and ambivalence in men's lifestyle magazines, *in* S. Henderson and A. Petersen (eds) *Consuming Health: The Commodification of Health Care.* London, Routledge, 187–203.

Burrows, R., Bunton, R., Muncer, S. and Gillen, K. (1995) The efficacy of health promotion, health economics and late modernism. *Health Education Research* 10: 241–249.

Bush, C. G. (1981) *In* American Association of University Women (1982). *Taking Hold of Technology: A Topic Guide for the 80s.* Washington DC, American Association of University Women.

Bush, C. G. (1983) Women and the assessment of technology: To think, to be; to unthink, to free, *in* J. Rothschild (ed.) *Machina ex Dea: Feminist Perspectives on Technology.* New York, Pergamon, 150–171.

Bush, J. (2000) It's just part of being a woman: Cervical screening, the body and femininity. *Social Science & Medicine* 50: 429–444.

Callon, M. and Latour, B. (1981) Unscrewing the Big Leviathan: How actors macro-structure reality and sociologists help them to do so, *in*

K. Knorr-Cetina and A. Cicourel (eds) *Advances in Social Theory and Methodology: Toward an Integration of Micro and Macro Sociologies*. London, Routledge.

Campbell, M. L. (1992) Nurses' professionalism in Canada: A labour process analysis. *International Journal of Health Services* 22: 751–765.

Canaan, J. E. (1996) One thing leads to another: Drinking, fighting and working class masculinities, *in* M. Mac an Ghaill (ed.) *Understanding Masculinities*. Buckingham, Open University Press, 114–125.

Canadian Institute for Health Information (2002) *Canada's Health Care Providers*. Ottawa, ON, Canadian Institute for Health Information.

Canadian Institute for Health Information (2006) *Health Personnel Trends in Canada, 1995 to 2004* (Revised July 2006). Ottawa, ON: Canadian Institute for Health Information.

Canadian Network for the Advancement of Research, Industry and Education (1997) *Towards a Canadian Health Iway: Vision, Opportunities and Future Steps*. Retrieved 6 February 2007 from: http://www.canarie.ca/funding/ehealth/health_reports.html.

Canadian Public Health Agency (2006) *What Determines Health*. Retrieved 19 November 2007 from: http://www.phac-aspc.gc.ca/ph-sp/phdd/determinants/index.html.

Canadian Standards Association (2003) *Coding of Work Injury or Disease Information*. Mississauga, ON, Canadian Standards Association.

Cancian, F. M. and Oliker, S. J. (2000) *Caring and Gender*. Walnut Creek, CA, Altamira Press.

Carvel, J. (2006, July 25) Hospitals' focus on waiting time targets led to 41 superbug deaths. *The Guardian*. Retrieved 8 June 2009 from: http://www.guardian.co.uk/society/2006/jul/25/hospitals.health.

Chaudhry, B., Wang, J., Wu, S., Maglione, M., Mojica, W., Roth, E., et al. (2006) Systematic review: Impact of health information technology on quality, efficiency, and costs of medical care. *Annals of Internal Medicine* 144: E-12–E-22.

Choiniere, J. (1993) A case study examination of nurses and patient information technology, *in* P. Armstrong, J. Choiniere and E. Day (eds) *Vital Signs: Nursing in Transition*. Toronto, ON, Garamond Press, 59–87.

Cloutier, E., David, H., Prévost, J. and Teiger, C. (1999) Importance of experience for older home care workers in facing up to the constraints of work. *Experimental Aging Research* 25: 405–410.

Cockburn, C. (1983) *Brothers: Male Dominance and Technological Change*. London, Pluto Press.

Cockburn, C. (1985) *Machinery of Dominance: Women, Men and Technical Know-How*. London, Pluto Press.

Cockburn, C. and Ormrod, S. (1993) *Gender and Technology in the Making*. London, Sage.

Cohen, M. G. and Cohen, M. (2004) *A Return to Wage Discrimination: Pay Equity Losses Through Privatization in Health Care*. Vancouver, BC, Canadian Centre for Policy Alternatives.

Cohen, M., Murphy, J., Nutland, K. and Ostry, A. (2005) *Continuing Care: Renewal or Retreat? BC Residential and Home Health Care Restructuring 2001–2004*. Vancouver, BC, Canadian Centre for Policy Alternatives.

Consalvo, M. and Paasonen, S. (2002) Introduction: On the Internet: Women matter, *in* M. Consalvo and S. Paasonen (eds) *Women and Everyday Uses of the Internet: Agency and Identity*. New York, Peter Lang, 1–18.

Cooper, J. and Weaver, K. D. (2003) *Gender and Computers: Understanding the Digital Divide*. Mahwah, NJ, Lawrence Erlbaum Associates.

Coppock, V., Haydon, D. and Richter, I. (1995) *The Illusions of 'Post-Feminism'*. London, Taylor & Francis.

Courtenay, W. H. (2000) Behavioural factors associated with disease: Injury and death among men: Evidence and implications for prevention. *Journal of Men's Studies* 9: 81–142.

Craig, D. and Brooks, R. (2006) *Plundering the Public Sector: How New Labour are Letting Consultants Run Off with £70 Billion of Our Money*. London, Constable.

Cranford, C. J., Vosko, L. F. and Zukewich, N. (2003) The gender of precarious employment in Canada. *Industrial Relations* 58: 454–482.

Curioso, W. H. (2006) New technologies and public health in developing countries: The Cell PREVEN Project, *in* M. Murero and R. E. Rice (eds) *The Internet and Health Care: Theory, Research and Practice*. Mahwah, NJ, Lawrence Erlbaum: 375–393.

Cwikel, J. M. and Israel, B. A. (1987) Examining mechanisms of social support and social networks: A review of health-related intervention studies. *Public Health Reviews* 15: 159–193.

Davis, K. and Kearley, M. (2005, March) *Personal Narratives from MacMorran Community Centre Concerning the Role of Community Access Computers in Individual and Community Capacity Development* (Available from the MacMorran Community Centre, P.O. Box 21046, St. John's, Newfoundland and Labrador A1A 5B2).

Daykin, N. and Naidoo, J. (1995) Feminist critiques of health promotion, *in* R. Bunton, S. Nettleton and R. Burrows (eds) *Sociology of Health Promotion: Critical Analyses of Consumption, Lifestyle and Risk*. London, Routledge.

Dejours, C. (1998a) 'Travailler' n'est pas 'déroger'. *Travailler* 1: 5–12.

Dejours, C. (1998b) *La Souffrance en France*. Paris, Éditions du Seuil.

Dembe, A. E. (1996) *Occupation and Disease: How Social Factors Affect the Conception of Work-Related Disorders*. New Heaven, Yale University Press.

Denton, M., Prus, S. and Walters, V. (2004) Gender differences in health: A Canadian study of the psychological, structural and behavioural determinants of health. *Social Science & Medicine* 58: 2585–2600.

Department of Finance and Administration (2005) *Foreword: Australians' Use and Satisfaction with E-Government Services*. Canberra, Australian Government Information Management Office.

Department of Health (2007) *Breast Screening: The Facts*. Retrieved 27 November 2007 from: http://www.cancerscreening.nhs.uk/breastscreen/publications/nhsbsp-the-facts-english-2007.pdf.

Department of Health and Ageing (2006) *Overview*. Retrieved 9 January 2007 from: http://www.health.gov.au/internet/wcms/publishing.nsf/Content/health-overview.htm.

Dorer, J. (2002) Internet and the construction of gender: Female professionals and the process of doing gender, *in* M. Consalvo and S. Paasonen (eds) *Women and Everyday Uses of the Internet: Agency and Identity*. New York, Peter Lang.

Doyal, L. (1995) *What Makes Women Sick: Gender and the Political Economy of Health*. New Brunswick, NJ, Rutgers University Press.

Doyal, L. (2001) Sex, gender, and health: The need for a new approach. *British Medical Journal* 323: 1061–1063.

Drudi, D. (1997) A century-long quest for meaningful and accurate occupational injury and illness statistics. *Compensation and Working Conditions Online*. Retrieved 31 January 2008 from: http://www.bls.gov/opub/cwc/archive/winter1997art3.pdf.

Durrance, J. C. (1989) Reference success: Does the 55% rule tell the whole story? *Library Journal* 114(April 15): 31–36.

Eaton, L. (2002) A third of Europeans and almost half of Americans use Internet for health information. *British Medical Journal* 325: 989.

Edwards, A. and Elwyn, G. (2001) Communicating and understanding risk. *Quality and Safety in Health Care* 10: i9–i13.

Eng, E. and Parker, E. (2002) Natural helper models to enhance a community's health and competence, *in* R. J. Diclemente, R. A. Crosby and M. C. Kegler (eds) *Emerging Theories in Health Promotion Practice and Research: Strategies for Improving Public Health*. San Francisco, Jossey-Bass, 126–156.

Ericksson-Zetterquist, U. (2007) Editorial: Gender and new technologies. *Gender, Work and Organizations* 14: 305–311.

Eysenbach, G. (2008) Medicine 2.0: Social networking, collaboration, participation, apomediation and openness. *Journal of Medical Internet Research* 10(3): e22.

Fallows, D. (2005) *How Women and Men Use the Internet*. Pew Internet and American Life Project, Retrieved 28 December 2005 from: www.pewtrusts. org/pdf/PIP_Women_Men_122805.pdf last visited 24 November 2006.

Faulkner, W. (2000) The power and the pleasure? A research agenda for 'making gender stick' to engineers. *Science, Technology & Human Values* 25: 87–119.

Faulkner, W. (2001) The technology question in feminism: A view from feminist technology studies. *Women's Studies International Forum* 24: 79–95.

Finkelstein, M. M. (1989) Analysis of mortality patterns and workers' compensation awards among asbestos insulation workers in Ontario. *American Journal of Industrial Medicine* 16: 523–528.

Fitzsimons, A. (2002) *Gender as Verb: Gender Segregation at Work*. Aldershot, Ashgate Publishing.

Fletcher, J. K. (1999) *Disappearing Acts: Gender, Power and Relational Practice at Work*. Cambridge, MIT Press.

Forster, P. (2005) *Queensland Health Systems Review*. Brisbane, The Consultancy Bureau and Queensland Government.

Fox, S. (2006) *Online Health Search 2006*. Pew Internet and American Life Project. Retrieved 31 January 2008 from: http://www.pewinternet.org/pdfs/PIP_Online_Health_2006.pdf.

Fujimura, J. (1987) Constructing 'do-able' problems in cancer research: Articulating alignment. *Social Studies of Science* 17: 257–293.

Gaby, S. and Henman, P. (2004) *E-Health: Transforming Doctor–Patient Relationships with a Dose of Technology*. Paper presented at the Australian National Governance Conference, Centre for Public Policy, University of Melbourne 14–15 April.

Gamber, W. (2003) Dressmaking, *in* N. E. Lerman, R. Oldenzeil and A. P. Mohun (eds) *Gender and Technology*. Baltimore, MD, Johns Hopkins University Press, 238–266.

Gannon, L. (1999) *Women and Aging: Transcending Myths*. New York, Routledge.

Gilbreth, F. (1912) *Primer of Scientific Management*. New York, Van Nostrand.

Gilligan, C. (1982) *In a Different Voice, Psychological Theory and Women's Development*. Cambridge, MA: Harvard University Press.

Gottlieb, B. H. (1981) Social networks and social support in community mental health, *in* B. H. Gottlieb (ed.) *Social Networks and Social Support*. Beverly Hills, Sage, 11–42.

Gøtzsche, P. and Nielsen, M. (2006) Screening for breast cancer with mammography. *Cochrane Database of Systematic Reviews* 2006(4): CD001877.

Graham, H. (1979) Prevention and health: Every mother's business, *in* C. Harris (ed.) *The Sociology of the Family: New Directions for Britain*. Sociological Review Monograph 28. Keele, University of Keele.

Graham, H. (1983) Caring: A labour of love, *in* J. Finch and D. Groves (eds) *A Labour of Love: Women, Work and Caring*. London: Routledge & Kegan Paul, 13–30.

Grant, K., Amaratunga, C., Armstrong, P., Boscoe, M., Pederson, A. and Willson, K. (eds) (2004) *Caring For/Caring About: Women, Home Care, and Unpaid Caregiving*. Aurora, ON, Garamond Press.

Green, E. (2001) Technology, leisure and everyday practices, *in* E. Green and A. Adam (eds) *Virtual Gender? Technology, Consumption and Identity*. London, Routledge, 173–188.

Green, E. and Adam, A. (1998) On-line leisure: Gender and ICTs in the home. *Information, Communication and Society* 1(3): 291–312.

Green, E. and Adam, A. (eds) (1999) Gender, society and ICTs [Special issue]. *Information, Communication & Society* 2(4): 399–586.

Green, E. and Adam, A. (eds) (2001) *Virtual Gender? Technology, Consumption and Identity*. London, Routledge.

Green, E., Griffiths, F., Henwood, F. and Wyatt, S. (2006) Desperately seeking certainty: Bone densitometry, the Internet and health care contexts, *in* A. Webster (ed.) *Innovative Health Technologies: New Perspectives, Challenge and Change*. Basingstoke, Palgrave Macmillan.

Green, E., Owen, J. and Pain, D. (eds) (1993) *Gendered by Design? Information Technology and Office Systems*. Basingstoke, Taylor & Francis.

Green, E. and Singleton, C. (2009) Mobile connections: An exploration of the place of mobile phones in friendship relations. *The Sociological Review* 57(1): 125–144.

Green, E. and Singleton, C. Gendering the digital, *in* N. Prior and K. Orson-Johnson (eds) *Re thinking Sociology in the Digital Age*. Palgrave MacMillan, (in press).

Green, E., Thompson, D. and Griffiths, F. (2002) Narratives of risk: Women at midlife, medical 'experts' and health technologies. *Health, Risk and Society* 4: 273–286.

Greenbaum, J. (2004) *Windows on the Workplace: Technology, Jobs, and the Organization of Office Work*, 3rd edn. New York, Monthly Review Press.

Greener, I. (2007) The politics of gender in the NHS: Impression management and 'getting things done.' *Gender, Work and Organizations* 14: 281–289.

Greer, G. (1991) *The Change: Women, Aging and the Menopause*. London, Hamish Hamilton.

Griffiths, F., Green, E. and Tsouroufli, M. (2005) The nature of medical evidence and its inherent uncertainty for the clinical consultation: Qualitative study. *British Medical Journal* 330: 511.

Griffiths, F., Lindenmeyer, A., Powell, J., Lowe, P. and Thorogood, M. (2006) Why are health care interventions delivered over the Internet? A systematic review of the published literature. *Journal of Medical Internet Research* 8(2): e-10.

Griffiths, S. (1996) Men's health: Unhealthy lifestyles and an unwillingness to seek medical help. *British Medical Journal* 312: 69–70.

Grint, K. and Gill, R. (1995) *The Gender-Technology Relation: Contemporary Theory and Research*. London, Taylor & Francis.

Grosz, E. (1994) *Volatile Bodies: Towards a Corporeal Feminism*. Bloomington, Indiana University Press.

Guillemin, M. (2004) Heart disease and mid-age women: Focusing on gender and age. *Health Sociology Review* 13: 7–14.

Guillemin, M. (2000) Blood, bone, women and HRT: Co-constructions in the menopause clinic. *Australian Feminist Studies* 15: 191–203.

Guillemin, M. (2001) Women and HRT: Working a position of critical compromise, *in* J. Daly, M. Guillemin and S. Hill (eds) *Technologies and Health: Critical Compromises*. Melbourne, Oxford University Press, 46–61.

Gullette, M. M. (2003) What to do when being aged by culture: Hidden narratives from the twentieth-century hormone debacle. *Generations* 27: 71–76.

Hacker, S. (1990) *'Doing It the Hard Way': Investigations of Gender and Technology*. Winchester, MA, Unwin Hyman.

Hankivsky, O. (2006) Beijing and beyond: Women's health and gender-based analysis in Canada. *International Journal of Health Services* 36: 377–400.

Haraway, D. (1985) A manifesto for cyborgs: Science, technology and socialist feminism in the 1980s. *Socialist Review* 80: 65–108.

Haraway, D. (1990) A manifesto for cyborgs: Science, technology and socialist feminism in the 1980s, *in* L. Nicholson (ed.) *Feminism/Postmodernism*. London, Routledge, 190–233.

Haraway, D. J. (1997) *Modest_Witness@Second_Millenium.FemaleMan©_Meets_Onco Mouse*™. New York, Routledge.

Hardey, M. (1999) Doctor in the house: The Internet as a source of lay health knowledge and the challenge to expertise. *Sociology of Health & Illness* 21: 820–835.

Hardey, M. (2008) Public health and Web 2.0. *Perspectives in Public Health* 128(4): 181–188.

Harding, S. (1986) *The Science Question in Feminism*. Milton Keynes, Open University Press.

Harding, S. (1997) Comment on Hekman's 'Truth and Method: Feminist Standpoint Theory Revisited': Whose standpoint needs the regimes of truth and reality? *Signs* 22: 382–391.

Hargiatti, E., DiMaggio, P. et al. (2004) Digital inequality: From unequal access to differentiated use, *in* K. Neckerman (ed.) *Social Inequality*. New York, Russell Sage Foundation, 355–400.

Harris, R. M. (1992) *Librarianship: The Erosion of a Woman's Profession*. Norwood, NJ, Ablex.

Harris, R. and Wathen, N. (2007) 'If my mother was alive I'd probably have called her.' Women's search for health information in rural Canada. *Reference & User Services Quarterly* 47: 67–79.

Harris, R. and Wilkinson, M. A. (2004) Situating gender: Students' perceptions of information work. *Information Technology & People* 17: 71–86.

Harris, R. M. and Dewdney, P. (1994) *Barriers to Information: How Formal Help Systems Fail Battered Women*. Westport, CT, Greenwood Press.

Harris, R. M., Wathen, C. N. and Fear, J. (2006) Searching for health information in rural Canada: Where do residents look for health information and what do they do when they find it? *Information Research* 12(1): paper 274.

Hartswood, M., Procter, R. N., Rouchy, P., Rouncefield, M., Slack, R. and Voss, A. (2003) Working IT out in medical practice: IT systems design and development as co-production, *in* ICT in Health Care: Sociotechnical Approaches, special issue, *Methods of Information in Medicine* 42: 392–397.

Hartswood, M., Proctor, R., Rouncefield, M., Slack, R., Soutter, J. and Voss, A. (2003) 'Repairing' the Machine: A case study of the evaluation of computer-aided detection tools in breast screening, *in* K. Kuutti, E. H. Karsten, G. Fitzpatrick, P. Dourish and K. Schmidt (eds) *Proceedings of the Eighth European Conference on Computer-Supported Cooperative Work*. Norwell, MA, Kluwer Academic Publishers, 375–394.

Harvey, D. (2006) *Spaces of Global Capitalism: Towards a Theory of Uneven Geographical Development*. London, Verso.

Hébert, F., Duguay, P. and Massicotte, P. (2003) *Les indicateurs de lésions indemnisées en santé et en sécurité du travail au Québec: Analyse par secteur d'activité économique en 1995–1997* (Report R-333). Montreal, QC, Institut de recherche Robert-Sauvé en santé et en sécurité du travail. Retrieved 31 January 2008 from: http://www.irsst.qc.ca/en/_publicationirsst_876.html.

Heisz, A. and LaRochelle-Côté, S. (2003) Working hours in Canada and the United States (Analytical Studies Research Paper Series, No. 11F0019 No.209). Ottawa, ON, Statistics Canada. Retrieved 31 January 2008 from: http://www.statcan.ca/english/research/11F0019MIE/11F0019MIE2003209.pdf.

Henderson, S. and Petersen, A. (2002) Introduction: Consumerism in health care, *in* S. Henderson and A. Petersen (eds) *Consuming Health: The Commodification of Health Care*. London, Routledge, 1–10.

Henman, M. J., Butow, P. N., Brown, R. F., Boyle, F. and Tattersall, M. H. N. (2002) Lay constructions of decision-making in cancer. *Psycho-Oncology* 11: 295–306.

Henwood, F. (1993) Establishing gender perspectives on information technology: Problems, issues and opportunities, *in* E. Green, J. Owen and D. Pain (eds) *Gendered By Design? Information Technology and Office Systems*. London, Taylor & Francis, 31–49.

Henwood, F. (1996) WISE choices? Understanding occupational decision-making in a climate of equal opportunities for women in science and technology. *Gender and Education* 8: 199–214.

Henwood, F. and Balka, E. (eds) (2005) e-Health. Special issue, *Information, Communication and Society* 7(4): 445–587.

Henwood, F., Harris, R., Burdett, S. and Marshall, A. (2008) Health intermediaries? Positioning the public library in e-health discourse, *in* C. N. Wathen, S. Wyatt and R. Harris (eds) *Mediating Health Information: The Go-Betweens in a Changing Socio-Technical Landscape.* Basingstoke, Palgrave MacMillan, 38–55.

Henwood, F. and Hart, A. (2003) Articulating gender in the context of ICTs in health care: The case of electronic patient records in the maternity services. *Critical Social Policy* 23: 249–267.

Henwood, F., Wyatt, S., Hart, A. and Smith, J. (2003) 'Ignorance is bliss sometimes': Constraints on the emergence of the 'informed patient' in the changing landscapes of health information. *Sociology of Health and Illness* 25: 589–607.

Herman, C. (1999) Women and the Internet, *in* Liberty (ed.) *Liberating Cyberspace: Civil Liberties, Human Rights and the Internet.* London, Pluto Press, 198–205.

Herman, C. and Webster, J. (2007) Gender and ICT [Special issue]. *Information, Communication & Society* 10(3): 279–432.

Hibbard, J. H., Greenlick, M., Jimison, H., Kunkel, L. and Tusler, M. (1999) Prevalence and predictors of the use of self-care resources. *Evaluation & The Health Professions* 22: 107–122.

Hirji, F. (2004) Freedom or folly? Canadians and the consumption of online health information. *Information, Communication & Society* 7: 445–465.

Hirokawa, K., Yagi, A. and Miyata, Y. (2002) Japanese social workers' healthy behaviours as related to masculinity: Focus on mental health workers and caregivers of children and nursing home residents. *International Journal of Psychology* 37: 353–359.

Howson, A. (1998) Embodied obligation: The female body and health surveillance, *in* S. Nettleton and J. Watson (eds) *The Body in Everyday Life.* London, Routledge, 218–240.

Howson, A. (2001) Locating uncertainties in cervical screening. *Health, Risk & Society* 3: 167–179.

Hugo, G. (2001, June) *Regional Australia: Definitions, Diversity and Dichotomy.* Paper presented to the Seminar program of Social Policy Research Centre, University of New South Wales, Sydney, NSW.

Huws, U. (2007) *Defragmenting: Towards a Critical Understanding of the New Global Division of Labor: Work Organisation Labour and Globalisation (Vol. 1, No. 2) (Work Organisation Labour and Globalisation).* Charleston, SC, BookSurge.

Huws, U. and Leys, C. (2003) *The Making of a Cyberteriat: Virtual Work in a Real World.* New York, Monthly Review Press.

Ison, T. (2005) *Recognition of Occupational Disease in Workers' Compensation. Canadian Centre for Occupational Health and Safety, Forum 2005*, March 3 and 4. Toronto, ON. Retrieved 31 January 2008 from: http://forum05.ccohs.ca/presentations/Terry_Ison_ENG_PDF.pdf.

Israel, B. A. (1985) Social networks and social support: Implications for natural helper and community level interventions. *Health Education Quarterly* 12: 65–80.

Israel, B. A. and Rounds K. A. (1987) Social networks and social support: A synthesis for health educators. *Advances in Health Education Promotion* 2: 311–351.

Jackson, B., Pederson, A. and Boscoe, M. (2006) *Gender-Based Analysis and Wait Times: New Questions, New Knowledge.* Toronto, Women and Health Care Reform Group. Retrieved 8 June 2009 from: http://www.womenandhealthcarereform.ca/publications/genderwaittimesen.pdf.

Johnson, C. A. (2004) Choosing people: The role of social capital in information seeking behaviour. *Information Research* 10(1): paper 201.

Jones, A., Henwood, F., Gart, A. and Gerhardt, C. (2003) Resistance at the frontline: The case of Electronic Patient Records (EPRs) in Maternity Services, *in* British Computer Society (ed.) *Healthcare Computing*. Guildford, British Computer Society.

Joyce, K. (2005) Appealing images: Magnetic resonance imaging and the production of authoritative knowledge. *Social Studies of Science* 35: 437–462.

Kenway, J. and Fitzclarence, L. (1997) Masculinity, violence and schooling: Controlling 'poisonous pedagogies'. *Gender & Education* 9: 117–134.

Kergoat, D. (1983) *Les ouvrières*. Paris, Sycomore.

Kickbusch, I., Wait, S. and Maag, D. (2006) *Navigating Health: The Role of Health Literacy*. London, International Longevity Centre-UK.

Kirby, M. G. (1999) Setting up a well man's clinic in primary care, *in* M. G. Kirby and R. N. Farah (eds) *Men's Health*. Oxford, Isis Medical Media, 321–344.

Kirkup, G., Janes, L., Woodward, K. and Hovenden, F. (2000) *The Gendered Cyborg: A Reader*. London, Routledge.

Kivits, J. (2004) Researching the 'informed patient': The case of online health information. *Information, Communication & Society* 7: 510–530.

Kohlberg, L. (1981) *Essays on Moral Development, Vol. 1, The Philosophy of Moral Development*. San Francisco, Harper and Row.

Kohlberg, L. (1984) *Essays on Moral Development, Vol. 2, The Psychology of Moral Development*. San Francisco, Harper and Row.

Kvasny, L. (2006) Cultural (re)production of digital inequality in a US community technology initiative. *Information, Communication & Society* 9: 160–181.

Langley, J. D. (1995) Experiences using New Zealand's hospital based surveillance system for injury prevention research. *Methods of Information in Medicine* 34: 340–344.

Larrabee, M. J. (1993) *An Ethic of Care: Feminist and Interdisciplinary Perspectives*. New York, Routledge.

Latour, B. (1992) The sociology of a few mundane artefacts, *in* W. Bijker and J. Law (eds) *Shaping Technology/Building Society*. Cambridge, MA, MIT Press.

Law, J. (1992) *Notes on the Theory of the Actor-Network: Ordering, Strategy and Heterogeneity*. Retrieved 8 June 2009 from Centre for Science Studies, Lancaster University Web site: http://www.lancs.ac.uk/fass/sociology/papers/law-notes-on-ant.pdf.

Lawson, H. M. (1999) Working on hair. *Qualitative Sociology* 22: 235–257.

Lee, S. Y. D., Arozullah, A. M. and Cho, Y. I. (2004) Health literacy, social support, and health: A research agenda. *Social Science & Medicine* 58: 1309–1321.

Le Jeune, G., Bélisle, A. C. and Messing, K. (2008) The data gap in Canadian women's occupational health. *Policy and Practice in Health and Safety* 6(2): 51–81.

Lerman, N. E., Oldenzeil, R. and Mohun, A. P. (2003) Introduction: Interrogating Boundaries, *in* N. E. Lerman, R. Oldenzeil and A. P. Mohun (eds) *Gender and Technology*. Baltimore, MD: Johns Hopkins University Press, 1–9.

Levinson, W., Kao, A., Kuby, A. and Thisted, R. A. (2005) Not all patients want to participate in decision making: A national study of public preferences. *Journal of General Internal Medicine* 20: 531–535.

Lie, M. (ed.) (2003) *He, She and IT Revisited: New Perspectives on Gender in the Information Society*. Oslo, Gyldendal Akademisk.

Lie, M. and Sørensen, K. (eds) (1996) *Making Technology Our Own, Domesticating Technology into Everyday Life*. Oslo, Scandinavian University Press.

Liff, S. and Shepherd, A. (2003) Gender and ICTs: New problems or a resolved issue? Paper presented at the Gender and ICTs Symposium, Brussels, January 20.

Light, J. (1999) When computers were women. *Technology & Culture* 40: 455–483.

Lippel, K. (1999) Workers' compensation and stress. Gender and access to compensation. *International Journal of Law and Psychiatry* 22: 79–89.

Lippel, K. (2003) Compensation for musculoskeletal disorders in Quebec: Systemic discrimination against women workers? *International Journal of Health Services* 33: 253–281.

Lippel, K. (2006a) L'expérience du processus d'appel en matière de lésions professionnelles telle que vécue par les travailleuses et les travailleurs, *in Développements récents en santé et sécurité du travail*, Vol 239. Cowansville, Éditions Yvon Blais, 119–180.

Lippel, K. (2006b) Precarious employment and occupational health and safety regulation in Quebec, *in* L. F. Vosko (ed.) *Precarious Employment: Understanding Labour Market Insecurity in Canada*. Montreal, QC and Kingston, ON, Mc-Gill Queen's University Press, 241–255.

Lloyd, T. (2001) Men and health: The context for practice, *in* N. Davidson and T. Lloyd (eds) *Promoting Men's Health: A Guide for Practitioners*. London, Balliere Tindall, 3–34.

Loader, B. and Keeble, L. (2004) *Challenging the Digital Divide? A Literature Review of Community Informatics Initiatives*. York, Joseph Rowntree Foundation. Retrieved 23 January 2008 from: http://www.jrf.org.uk/bookshop/eBooks/1859351980.pdf.

Lock, M. (1998) Anomalous ageing: Managing the postmenopausal body. *Body and Society* 4: 35–61.

Lupton, D. (1994) Femininity, responsibility, and the technological imperative: Discourses on breast cancer in the Australian press. *International Journal of Health Services* 24: 73–89.

Lupton, D. (1995) *The Imperative of Health: Public Health and the Regulated Body*. London, Sage Publications.

Lupton, D. (1997) Consumerism, reflexivity and the medical encounter. *Social Science & Medicine* 45: 373–381.

MacKenzie, D. and Wajcman, J. (eds) (1984) *The Social Shaping of Technology*. Milton Keynes, Open University Press.

MacKenzie, D. and Wajcman, J. (eds) (1999) *The Social Shaping of Technology*, 2nd Edn. Milton Keynes, Open University Press.

Mackian, S., Bedri, N. and Lovel, H. (2004) Up the garden path and over the edge: Where might health-seeking behaviour take us? *Health Policy and Planning* 19: 137–146.

Madden, M. and Fox, S. (2006) *Finding Answers Online in Sickness and in Health*. Washington, DC: Pew Internet & American Life Project. Retrieved 5 July 2006 from: http://pewresearch.org/reports/?ReportID=20.

Malone, R. E. (2003) Distal nursing. *Social Science and Medicine* 56: 2317–2326.

Marshall, A. (2004) ICTs for health promotion in the community: A participative approach, *in* P. Day and D. Schuler (eds) *Community Practice in the Network Society: Local Action/Global Interaction*. London, Routledge, 78–91.

Marshall, J. G. (1992) The impact of the hospital library on clinical decision making: The Rochester study. *Bulletin of the Medical Library Association* 80: 169–178.

Martin, E. (1987) *The Woman in the Body*. Milton Keynes, Open University Press.

McDiarmid, M. A. and Gucer, P. W. (2001) The 'GRAS' status of women's work. *Journal of Occupational and Environmental Medicine* 43: 665–669.

McNeil, M. (ed.) (1987) *Gender and Expertise*. London, Free Association Press.

McPherson, K. (1996) *Bedside Matters: The Transformation of Canadian Nursing, 1900–1990*. Toronto, ON, Oxford University Press.

Meadows, L. M., Thurston, W. E. and Berenson, C. A. (2001) Health promotion and preventive measures: Interpreting messages at midlife. *Qualitative Health Research* 11: 450–463.

Mehlum, I. S., Kjuus, H., Veiersted, K. B. and Wergeland, E. (2006) Self-reported work-related health problems from the Oslo health study. *Occupational Medicine* 56: 371–379.

Messing, K., Courville, J., Boucher, M., Dumais, L. and Seifert, A. M. (1994) Can safety risks of blue-collar jobs be compared by gender? *Safety Science* 18: 95–112.

Messing, K. (1998a) Hospital trash: Cleaners speak of their role in disease prevention. *Medical Anthropology Quarterly* 12: 168–187.

Messing, K. (1998b) *One-Eyed Science: Occupational Health and Women Workers*. Philadelphia, PA, Temple University Press.

Messing, K. (2002) La place des femmes dans les priorités de recherche en santé au travail au québec. *Industrial Relations/Relations industrielles* 57: 660–686.

Messing, K. and Stellman, J. M. (2006) Sex, gender and women's occupational health: The importance of considering mechanism. *Environmental Research* 101: 149–162.

Messing, K. and Boutin, S. (1997) La reconnaissance des conditions difficiles dans les emplois des femmes et les instances gouvernementales en santé et en sécurité du travail. *Relations Industrielles/Industrial Relations* 52: 333–362.

Messing, K., Chatigny, C. and Courville, J. (1998) 'Light' and 'heavy' work in the housekeeping services of a hospital. *Applied Ergonomics* 29: 451–459.

Miller, D. and Slater, D. (2000) *The Internet: An Ethnographic Approach*. Oxford, Berg.

Mills, A. J. and Tancred, P. (eds) (1992) *Gendering Organizational Analysis*. London, Sage, 1–7.

Mimoto, H. and Cross, P. (1991) The growth of the federal debt. *The Canadian Economic Observer* 4(6): 1–17.

Mitchell, R. E. and Hurley, D. J. (1981) Collaboration with natural helping networks: Lessons from studying paraprofessionals, *in* B. H. Gottlieb (ed.) *Social Networks and Social Support*. Beverly Hills, CA, Sage, 277–298.

Moen, A. and Brennan, P. F. (2005) Health@Home: The work of health information management in the household (HIMH): Implications for consumer health informatics (CHI) innovations. *Journal of the American Medical Informatics Association* 12: 648–656.

Moon, J. and Fisher, J. (2006) *The Effectiveness of Australian Medical Portals: Are They Meeting the Health Consumers' Needs?* Paper presented at the 19th Bled eConference, eValues, Bled, Slovenia, 5–7 June.

Morris, M. (2001) *Gender-Sensitive Home and Community Care and Caregiving Research: A Synthesis Paper.* Ottawa, ON, Women's Health Bureau, Health Canada.

Moyal, A. (1992) The gendered use of the telephone: An Australian case study. *Media, Culture & Society* 14: 51–72.

Mühlhauser, I. and Berger, M. (2000) Evidence-based patient information in diabetes. *Diabetic Medicine* 17: 823–829.

Murero, M. and Rice, R. E. (2006) *The Internet and Health Care: Theory, Research and Practice.* Mahwah, NJ, Lawrence Erlbaum.

National Health Information Management Advisory Council (2001) *Health Online: A Health Information Action Plan for Australian.* Canberra, Department of Health and Aged Care.

National Rural Health Policy Sub-committee, & National Rural Health Alliance (2002) *Healthy Horizons: A Framework for Improving the Health of Rural, Regional and Remote Australians 2003–2007.* Canberra, Australian Health Ministers Conference.

Nettleton, S. and Burrows, R. (2003) E-scaped medicine? Information, reflexivity and health. *Critical Social Policy* 23: 165–185.

Nettleton, S., Burrows, R. and O'Malley, R. (2005) The mundane realities of the everyday lay use of the Internet for health, and their consequences for media convergence. *Sociology of Health & Illness* 27: 972–992.

Nettleton, S., Burrows, R., O'Malley, L. and Watt, I. (2004) Health E-Types? An analysis of the everyday use of the Internet for health. *Information, Communication and Society* 7: 531–553.

Neysmith, S. M. (2000) Networking across difference: Connecting restructuring and caring labour, *in* S. M. Neysmith (ed.) *Restructuring Caring Labour: Discourse, State Practice, and Everyday Life.* Don Mills, ON, Oxford University Press, 1–28.

NHS Executive (1998) *National Survey of NHS Patients, General Practice.* London, Stationary Office.

Nicholas, D., Huntington, P., Jamali, H. and Williams, P. (2007) *Digital Health Information for the Consumer: Evidence and Policy Implications.* Aldershot, Ashgate.

Nilsen, K. and McKechnie, E. F. (2002) Behind closed doors: An exploratory study of the perceptions of librarians and the hidden intellectual work of collection development. *Library Quarterly* 72: 294–325.

Novek, J. (2002) IT, gender, and professional practice: Or, why an automated drug distribution system was sent back to the manufacturer. *Science, Technology, and Human Values* 27: 379–403.

O'Cathain, A., Sampson, F. C., Munro, J. F., Thomas, K. J. and Nicholl, J. P. (2004) Nurses' views of using computerized decision software in NHS Direct. *Journal of Advanced Nursing* 45: 280–286.

Olerup, A., Schneider, L. and Monod, E. (eds) (1985) *Women Work and Computerization: Opportunities and Advantages.* Amsterdam, Elsevier.

O'Neill, O. (2002) *A Question of Trust: The BBC Reith Lectures 2002.* Cambridge, Cambridge University Press.

Ordre des psychologues du Québec (2002) La vie au travail: Un monde en transformation. *Symposium Santé mentale au travail*. Montréal, QC, Ordre des Psychologues du Québec.

Orlikowski, W. (1992) The duality of technology: Rethinking the concept of technology in organisations. *Organisational Science* 3: 398–427.

Ormrod, S. (1994) 'Let's nuke the dinner': Discursive practices of gender in the creation of a new cooking process, *in* C. Cockburn and R. Furst-Dilic (eds) *Bringing Technology Home: Gender and Technology in a Changing Europe*. Buckingham, Open University Press, 42–58.

Ormrod, S. (1995) Feminist sociology and methodology: Leaky black boxes in gender/technology relations, *in* K. Grint and R. Gill (eds) *The Gender-Technology Relation: Contemporary Theory and Research*. London, Taylor & Francis, 31–47.

Ostry, A. (2006) *Change and Continuity in Canada's Health Care System*. Ottawa, ON, Canadian Healthcare Association Press.

Ouellet, F. (2003) *La SST: Un système détourné de sa mission*. Napierville, QC, Sansectra/Saint Lambert, QC, Impact.

Paechter, C. (2006) Masculine femininities/feminine masculinities: Power, identities and gender. *Gender and Education* 18: 253–263.

Pandey, S. K., Hart, J. J. and Tiwary, S. (2003) Women's health and the Internet: Understanding emerging trends and implications. *Social Science & Medicine* 56: 179–191.

Pettigrew, K. E. (2000) Lay information provision in community settings: How community health nurses disseminate human services information to the elderly. *Library Quarterly* 70: 47–85.

Phillips, A. and Talyor, B. (1980) Sex and skill: Notes towards a feminist economics. *Feminist Review* 6: 79–88.

Pickstone, J. (2000) *Ways of Knowing: A New Science, Technology and Medicine*. Manchester, Manchester University Press.

Plant, S. (1997) *Zeros + Ones: Digital Women and the New Technoculture*. London, Fourth Estate.

Policy (2008, September 13) *Wikipedia*. Retrieved 12: 37, 18 November 2007 from: http://en.wikipedia.org/wiki/Policies.

Poole, M. and Isaacs, D. (1997) Caring: A gendered concept. *Women's Studies International Forum* 4: 529–536.

Porter, T. M. (1995) *Trusts in Numbers: The Pursuit of Objectivity in Science and Public Life*. Princeton, NJ, Princeton University Press.

Postl, B. (2006) *Final Report of the Federal Advisor on Wait Times*. Ottawa, ON, Health Canada.

Pransky, G. S., Snyder, T., Dembe, A. and Himmelstein, J. (1999) Under-reporting of work-related disorders in the workplace: A case study and review of the literature. *Ergonomics* 42: 171–182.

Pratt, W., Reddy, M. C., McDonald, D. W., Tarczy-Hornoch, P. and Gennari, J. H. (2004) Incorporating ideas from computer-supported cooperative work. *Journal of Biomedical Informatics* 37: 128–137.

Preece, J. and Ghozati, K. (2001) Experiencing empathy online, *in* R. Rice and J. Katz (eds) *The Internet and Health Communication: Experiences and Expectations*. London, Sage Publications, 237–260.

Public Health Information Development Unit (2001) *ASGC Remoteness Classification, Australia*. Adelaide, Public Health Information Development Unit.

Rachlis, M. (2005) Medicare: Innovation is the key to sustainability? *Healthcare Management Forum* 18: 28–31.

Ramasubbu, K., Gurm, H. and Litaker, D. (2001) Gender bias in clinical trials: Do double standards still apply? *Journal of Women's Health and Gender-Based Medicine* 10: 757–764.

Reventlow, S. (2002) From accident to diagnosis, cultural responses to the risk of osteoporosis. *Revista della Societa Italiana di Antropologia Medica Italiana* 13–14: 87–99.

Reventlow, S. and Bang, H. (2006) Brittle bones: Ageing or threat of disease? Exploring women's cultural models of osteoporosis. *Scand J Public Health.* 34(3):320–326.

Reventlow, S. D., Hvas, L. and Malterud, K. (2006) Making the invisible body visible. Bone scans, osteoporosis and women's bodily experiences. *Social Science & Medicine* 62: 2720–2731.

Reverby, S. M. (1987) *Ordered to Care: The Dilemma of American Nursing, 1850–1945.* New York, Cambridge University Press.

Richardson, J. C., Hassell, A. B., Hay, E. M. and Thomas, E. (2002) 'I'd rather go and know': Women's understanding and experience of DEXA scanning for osteoporosis. *Health Expectations* 5: 114–126.

Rogers, M. and Todd, C. (2002) Information exchange in oncology outpatient clinics: Source, valence and uncertainty. *Psycho-Oncology* 11: 336–345.

Ross, C. S. and Dewdney, P. (1994) Best practices: An analysis of the best (and worst) in fifty-two public library reference transactions. *Public Libraries* 33: 261–266.

Rothschild, J. (ed.) (1983) *Machina ex Dea: Feminist Perspectives on Technology.* New York, Pergamon.

Royal College of Physicians and Bone and Teeth Society of Great Britain (2001) *Osteoporosis: Clinical Guidelines for Prevention and Treatment.* London, Royal College of Physicians.

Sackett, D. L., Rosenberg, W. M., Gray, J. A., Haynes, R. B. and Richardson, W. S. (1996) Evidence based medicine: What it is and what it is not. *British Medical Journal* 312: 71–72.

Salander, P. and Henriksson, R. (2005) Severely diseased lung cancer patients narrate the importance of being included in a helping relationship. *Lung Cancer* 50: 155–162.

Sandelowski, M. (2000) *Devices and Desires: Gender, Technology and American Nursing.* Chapel Hill, The University of North Carolina Press.

Schmidt, K. and Wagner, I. (2004) Ordering systems: Coordinative practices and artifacts in architectural design and planning. *Computer Supported Cooperative Work* 5–6: 349–408.

Schofield, T., Connell, R., Walker, L., Wood, J. and Butland, D. (2000) Understanding men's health and illness: A gender-relations approach to policy, research and practice. *Journal of American College Health* 48: 247–258.

Scott, A., Semmens, L. and Willoughby, L. (2001) Women and the Internet: The natural history of a research project, *in* E. Green and A. Adam (eds) *Virtual Gender: Technology, Consumption and Identity.* London, Routledge, 3–22.

Scott, C. M. and Thurston, W. E. (2004) The influence of social context on partnerships in Canadian health systems. *Gender, Work and Organizations* 11: 481–505.

Seidler, V. (1998) Masculinity, violence and emotional life, *in* G. B. S. Williams (ed.) *Emotions in Social Life*. London, Routledge, 193–210.

Seifert, A. M. and Messing, K. (2004) Looking and listening in a technical world: Effects of discontinuity in work schedules on nurses' work activity. *Perspectives interdisciplinaires sur le travail et la santé* 6(1): 1–15.

Seifert, A. M., Messing, K. and Elabidi, D. (1999) Analyse des communications et du travail des préposées à l'accueil d'un hospital pendant la restructuration des services. *Recherches féministes* 12: 85–108.

Selwyn, N. (2003) Apart from technology: Understanding people's nonuse of information and communications technologies in everyday life. *Technology in Society* 25: 99–116.

Shannon, H. S. and G. S. Lowe (2002) How many injured workers do not file claims for workers' compensation benefits? *American Journal of Industrial Medicine* 42: 467–473.

Shumaker, S. and Hill, D. R. (1991) Gender differences in social support and physical health. *Health Psychology* 10: 102–111.

Sievert, M. C., Patrick, T. B. and Reid, J. C. (2001) Need a bloody nose be a nosebleed? or, lexical variants cause surprising results. *Bulletin of the Medical Library Association* 89: 68–71.

Silverstone, R. and Hirsch, E. (eds) (1992) *Consuming Technologies, Media and Information in Everyday Spaces*. London, Routledge.

Simpson, L., Hall, M. and Leggett, S. (2006) 'We're all out there busting our guts, trying to do the best we can for our people': Health intermediaries in Australian indigenous communities, *in* C. N. Wathen, S. Wyatt and R. Harris (eds) *Mediating Health Information: The Go-Betweens in a Changing Socio-Technical Landscape*. Basingstoke, Palgrave MacMillan, 150–166.

Simpson, L., Stockwell, M., Leggett, S., Wood, L. and Penn, D. (2006) A capacity building approach to health literacy through ICTs, *in* A. A. Lazakidou (ed.) *Handbook of Research on Informatics in Healthcare and Biomedicine*. Hershey, PA, Idea Group, 431–437.

Singleton, V. (1995) Networking constructions of gender and constructing gender networks: Considering definitions of women in the British cervical screening programme, *in* K. Grint and R. Gill (eds) *The Gender-Technology Relation: Contemporary Theory and Research*. London, Taylor & Francis, 146–173.

Skeggs, B. (1997) *Formations of Class and Gender: Becoming Respectable*. London, Sage.

Smith, D. (2002) Institutional ethnography, *in* T. May (ed.) *Qualitative Research: An International Guide to Issues in Practice*. London: Sage, 17–52.

Spender, D. (1995) *Nattering on the Net: Women, Power and Cyberspace*. Toronto, ON, Garamond Press.

Spink, A., Yang, Y., Janse, J., Nykanen, P., Lorence, D. P., Ozmutlu, S. and Ozmutlu, H. C. (2004) A study of medical and health queries to web search engines. *Health Information and Libraries Journal* 21: 44–51.

Spitzer, R. L. (1981) The diagnostic status of homosexuality in DSM-III: A reformulation of issues. *American Journal of Psychiatry* 138: 210–215.

Star, S. L. (1991) Invisible work and silenced dialogues in knowledge representation, *in* I. Eriksson, B. Kitchenham and K. Tidjens (eds) *Women, Work and Computerization*. Amsterdam, North Holland, 81–92.

Star, S. L. and Ruhleder, K. (1996) Steps toward an ecology of infrastructure: Design and access for large information spaces. *Information Systems Research* 7: 111–134.

Star, S. L. and Strauss, A. (1999) Layers of silence, arenas of voice: The ecology of visible and invisible work. *Computer Supported Cooperative Work* 8: 9–30.

Statistics Canada (2005) Labour force and participation rates by sex and age group. *CANSIM*, Table 282-0002. Retrieved 8 June 2009 from: http: //cansim2. statcan.ca/cgi-win/cnsmcgi.exe?Lang=E&RootDir=CII/& ResultTemplate=CII/ CII_&Array_Pick=1&ArrayId=2820002.

Statistics Canada (2006, July 19) General social survey: Paid and unpaid work. *The Daily*. Retrieved 1 October 2007 from: http://www.statcan.ca/Daily/ English/060719/d060719b.htm.

Status of Women Canada (1998) *Gender-Based Analysis: A Guide for Policy-Making*. Retrieved 23 September 2008 from: http://www.swc-cfc.gc.ca/pubs/ gbaguide/gbaguide_e.html.

Stein, J. (2001) *The Cult of Efficiency*. Toronto, ON, House of Anansi Press.

Steinberg, R. and Haignere, L. (1987) Equitable compensation: Methodological criteria for comparable worth, *in* C. Bose and G. Spitze (eds) *Ingredients for Women's Employment Policy*. Albany, NY, State University of New York Press, 157–82.

Stock, S., Tissot, F., Messing, K. and Goudreau, S. (2004) Can 1998 Quebec Health Survey data help us estimate underreporting of Workers' Compensation lost-time claims for musculoskeletal disorders of the neck, back and upper extremity? *Proceedings of the 4th International PREMUS Conference* 2: 573–574.

Stoller, E. P. (1993) Gender and the organization of lay health care: A socialist-feminist perspective. *Journal of Aging Studies* 7: 151–170.

Stone, A. R. (1995) *The War of Desire and Technology at the Close of the Mechanical Age*. Cambridge, MIT Press.

Struthers, J. (1987) Lord give us men: Women and social work in English Canada, 1918–1953, *in* A. Moscovitch and J. Albert (eds) *The Benevolent State: The Growth of Welfare in Canada*. Toronto, ON, Garamond Press, 111–125.

Suchman, L. (1983) Office procedures as practical action: Models of work and system design. *ACM Transactions on Office Information Systems* 1: 320–328.

Suchman, L. (2002) Located accountabilities in technology production. *Scandinavian Journal of Information Systems* 14: 91–105.

Suchman, L. and Jordan, B. (1989) Computerization and women's knowledge, *in* K. Tijdens (ed.) *Women, Work and Computerization*. Amsterdam, North Holland, 153–160.

Suitor, J. J. and Pillemer, K. (2002) Gender, social support, and experiential similarity during chronic stress: The case of family caregivers, *in* J. A. Levy and B. A. Pescosolido (eds) *Social Networks and Health*. Oxford, Elsevier Science, 247–266.

Teiger, C. and Plaisantin M. C. (1984) Les contraintes du travail dans les travaux répétitifs de masse et leurs conséquences sur les travailleuses, *in* J. A. Bouchard (ed.) *Les effets des conditions de travail sur la santé des travailleuses*. Montreal, QC, Confédération des Syndicats Nationaux, 33–68.

Teiger, C. and Bernier, C. (1990) Intérêt de l'analyse ergonomique du travail pour mettre en évidence les compétences méconnues: Le cas des tâches de saisie dans le tertiaire informatisé, *in* C. Brabant and K. Messing (eds) *'Sexe faible' ou travail ardu?* Montreal, Association Canadienne-Française pour l'Avancement des Sciences, 61–70.

Teiger, C. and Bernier, C. (1992) Ergonomic analysis of work activity of data entry clerks in the computerized service sector can reveal unrecognized skills. *Women and Health* 18(3): 67–77.

Tellioglu, H. and Wagner, I. (2001) Work practice surrounding PACS: The politics of space in hospitals. *Computer Supported Cooperative Work* 10: 163–188.

Thomlinson, E., McDonagh, M. K., Crooks, K. B. and Lees, M. (2004) Health beliefs of rural Canadians: Implications for practice. *Australian Journal of Rural Health* 12: 258–263.

Thompson, E. P. (1978) *The Poverty of Theory and Other Essays*. London, Merlin Press.

Timmons, S. (2003a) Nurses resisting information technology. *Nursing Inquiry* 10: 257–269.

Timmons, S. (2003b) A failed panopticon: Surveillance of nursing practice via new technology. *New Technology, Work and Employment* 18: 143–153.

Treichler, P., Cartwright, L. and Penley, C. (eds) (1998) *The Visible Woman: Imaging Technologies, Gender and Science*. New York, New York University Press.

Tronto, J. C. (1993) *Moral Boundaries: A Political Argument for an Ethics of Care*. New York and London, Routledge.

Tuominen, K. (2004) 'Whoever increases his knowledge merely increases his heartache.' Moral tensions in heart surgery patients' and their spouses' talk about information seeking. *Information Research* 10(1): paper 202.

Turkle, S. (1984) *The Second Self: Computers and the Human Spirit*. New York, Simon and Schuster.

Turkle, S. (1995) *Life on the Screen: Identity in the Age of the Internet*. New York, Simon and Schuster.

Van Eerd, D., Côté, P., Beaton, D., Hogg-Johnson, S., Vidmar, M. and Kristman, V. (2006) Capturing cases in workers' compensation databases: The example of neck pain. *American Journal of Industrial Medicine* 49: 557–568.

Vézina, M. and Bourbonnais, R. (2001) Incapacité de travail pour des raisons de santé mentale, in *Portrait social du Québec: Données et analyses*. Quebec, Institut de la statistique du Québec, 279–287.

Vickers, J. (1984) *Taking Sex into Account: The Policy Consequences of Sexist Research*. Ottawa, ON, Carleton University Press.

Vlassoff, C. and Moreno, C. G. (2002) Placing gender at the centre of health programming: Challenges and limitations. *Social Science and Medicine* 54: 1713–1723.

Vosko, L. F. (2006) Precarious employment: Towards an improved understanding of labour market insecurity, *in* L. F. Vosko (ed.) *Precarious Employment: Understanding Labour Market Insecurity in Canada*. Montreal, QC and Kingston, ON, McGill-Queen's University Press, 3–39.

Wagner, I. (1993) Women's voice: The case of nursing information systems. *AI and Society* 7: 295–310.

Wajcman, J. (1991) *Feminism Confronts Technology*. University Park, Pennsylvania State University Press.

Wajcman, J. (2004) *TechnoFeminism*. Cambridge, Polity Press.

Wajcman, J. (2006) New connections: Social studies of science and technology and studies of work. *Work, Employment and Society* 20: 773–786.

Wajcman, J. (2007) From women and technology to gendered technoscience. *Information, Communication & Society* 10: 287–298.

Wakeford, N. (1997) Networking women and grrrls with information/communication technology: Surfing tales on the World Wide Web, *in* J. Terry and M. Calvert (eds) *Processed Lives: Gender and Technology in Everyday Life.* London, Routledge, 51–66.

Warner, D. and Procaccino, J. D. (2003) Toward wellness: Women seeking health information. *Journal of the American Society for Information Science and Technology* 55: 709–730.

Wathen, C. N., Wyatt, S. and Harris, R. (eds) (2008) *Mediating Health Information: The Go-Betweens in a Changing Socio-Technical Landscape.* Basingstoke: Palgrave Macmillan.

Webb, G. R., Redman, S., Wilkinson, C. and Sanson-Fisher, R. W. (1989) Filtering effects in reporting work injuries. *Accident Analysis and Prevention* 21: 115–123.

WebMD University. *Mom: The Nutritional Gatekeeper.* Retrieved 11 December 2006 from: www.webmd.com/content/pages/18/101909.htm.

Webster, A. (2002) Innovative health technologies and the social: Redefining health, medicine and the body. *Current Sociology* 50: 443–457.

Webster, A. (ed.) (2006) *New Technologies in Health Care: Challenge, Change and Innovation.* Basingstoke, Palgrave Macmillan.

Webster, A. (2007) *Health, Technology and Society: A Sociological Critique.* Basingstoke, Palgrave Macmillan.

Webster, A. and Brown, N. (2004) *New Medical Technologies and Society: Reordering Life.* Cambridge, Polity Press.

Webster, J. (1996) *Shaping Women's Work: Gender, Employment and Information Technology.* White Plains, NY, Longman.

Wellman, B. and Wortley, S. (1990) Different strokes from different folks: Community ties and social support. *American Journal of Sociology* 96: 558–588.

White, J. (1988) Women in leisure service management, *in* E. Wimbush and M. Talbot (eds) *Relative Freedoms: Women and Leisure.* Milton Keynes: Open University Press, 147–160.

White, J. (2003) Changing labour process and the nursing crisis in Canadian hospitals, *in* C. Andrew, P. Armstrong, H. Armstrong, W. Clement and L. F. Vosko (eds) *Studies in Political Economy: Developments in Feminism.* Toronto, ON, Women's Press, 125–154.

Wilkin, T., Devendra, D., Dequeker, J. and Luyten, F. P. (2001) Education and debate: Bone densitometry is not a good predictor of hip fracture. *British Medical Journal* 323: 795–799.

Williams, S. J., Birke, L. and Bendelow, G. (2003) *Debating Biology: Sociological Reflections on Health, Medicine and Society.* London, Routledge.

Willis, K. and Baxter, J. (2003) Trusting technology: Women aged 40–49 years participating in screening for breast cancer—an exploratory study. *Australian and New Zealand Journal of Public Health* 27: 282–286.

Wilson, R. and Hubert, J. (2002) Resurfacing the care in nursing by telephone: Lessons from ambulatory oncology. *Nursing Outlook* 50: 160–164.

Witz, A. (1992) *Professions and Patriarchy.* London, Routledge.

Wolff, J. (1977) Women in organizations, *in* S. Clegg and D. Dunkerley (eds) *Critical Issues in Organizations*. London, Routledge & Kegan Paul, 7–20.

Woodfield, R. (2000) *Women, Work and Computing*. Cambridge, Cambridge University Press.

Wright, E. B., Holcombe, C. and Salmon, P. (2004) Doctors' communication of trust, care, and respect in breast cancer: Qualitative study. *British Medical Journal* 328: 864, doi: 10.1136/bmj.38046.771308.7C (published 30 March 2004).

Wuest, J. (2000) Negotiating with helping systems: An example of grounded theory evolving through emergent fit. *Qualitative Health Research* 10: 51–70.

Wyatt, S. (2008) Feminism, technology and the information society: Learning from the past, imagining the future. *Information, Communication and Society* 11(1): 111–130.

Wyatt, S., Henwood, F., Hart, A. and Smith, J. (2005) The digital divide, health information and everyday life. *New Media & Society* 7: 199–218.

Wyatt, S., Thomas, T. and Terranova, T. (2002) They came, they surfed, they went back to the beach: Conceptualising use and non-use of the internet, *in* S. Woolgar (ed.) *Virtual Society? Technology, Cyberbole, Reality*. Oxford, Oxford University Press, 23–40.

Ziebland, S. (2004) The importance of being expert: The quest for cancer information on the Internet. *Social Science & Medicine* 59: 1783–1793.

Zimmerman, M. K., Litt, J. S. and Bose, C. E. (eds) (2006) *Global Dimensions of Gender and Carework*. Stanford, CA, Stanford University Press.

Index